Brain, Mind, and Spirit

Brain, Mind, and Spirit

Lloyd Vinnedge

VANTAGE PRESS
New York / Washington / Atlanta
Los Angeles / Chicago

FIRST EDITION

Published by Vantage Press, Inc.
516 West 34th Street, New York, New York 10001

Manufactured in the United States of America
ISBN: 0-533-07097-X

Library of Congress Catalog Card No.: 86-90182

CONTENTS

INTRODUCTION

My interest in the human brain came about when I took a course in physiological psychology. I had read in the field of psychology for years, but I had not read much about the brain, since most of my reading was concerned with counseling and psychotherapy. I had focused most of my thinking upon emotions and how they affected us as human beings. I had read much of Freud's work, and his emphasis upon the unconscious was especially interesting to me, as it still is today. This I take to be one of Freud's significant contributions to the understanding of our mental processes: He opened up a whole new viewpoint in dealing with the emotional life. Now we have to consider the unconscious as well as those things of which we are conscious.

My interest in the New Testament began even before my college days and has continued to the present. So I have had for years what might be called a double intellectual interest. This has caused me to look at religious matters largely from a psychological point of view and to look at much of psychology from a religious point of view. Not everyone can or should do this, but for me it turned out to be a necessity, since I could not let go of either field of study. Until I began my study of the brain, I gave little credence to the effect that the brain and the rest of human biology had upon our behavior as humans. I have had to revise my thinking about such matters and realize just how much effect the brain has upon behavior. It seems to me that we have learned quite a lot about the brain, the way it works, and the effects it has on human behavior, but it seems to me also that we still have a great deal to learn about the brain and how it works. This is certainly a promising field for research and for exploration; it is also as important as it is promising. Such study could, for example, help us understand much about criminal behavior, especially behavior that seems totally unacceptable in modern society. Part of the price we will have to pay for such advances is to admit just how much of primitive animal behavior is still in us. We like to think of ourselves as refined, sophisticated, urbane, polite, generous, and perhaps even magnanimous. And to a degree we surely are. But when we consider the inhuman things that we do to each other from time to time, how much money and time our society spends trying to solve murders and other crimes of all sorts, then we surely must admit that something is amiss in ourselves. What is there in ourselves that allows us to do such generous things on the one hand and yet commit such flagrant crimes on the other? I am convinced that at least part of the answer lies right within our brain.

This work is all too inadequate; I am the first to state that. It is just a small beginning of what I hope will become a full-blown field of study in the future. If that happens, it will be because people much better qualified than I am will pick up the strands and pursue them wherever they lead. If this happens, I am convinced that all mankind will benefit.

Brain, Mind,
and
Spirit

Chapter One
THE MARVELOUS BRAIN

At the crack of the bat, an outfielder turns and runs toward the fence and at the last second turns and makes a spectacular catch of the baseball. A quarterback takes the snap from the center and drops back to pass. At the same time, opposing linesmen are coming at him, but just before he is hit, he throws the ball in a perfect strike to his tight end for a touchdown. A man is driving along in his car on a two-lane highway and looks ahead of the car in front of him; the road appears to be clear. But just as he gets alongside the car he is passing, an oncoming car appears just over a slight hill in front of him. His heart starts to speed up, but at the same time he realizes that he has time to pass the car and get back safely in his own lane. His heartbeat quickly returns to normal, and he goes on his way to his appointment.

All of these examples are commonplace, and we hardly think anything of them. Yet when we stop and think about them, they are somewhat of a marvel. How does the outfielder know whether to drop back or run in at the crack of the bat against the ball? Experience helps, good hearing and eyesight help, but it is the marvelous brain and nervous system that we humans possess that make such feats possible. The brain must process in an instant what the ear hears, what the eye sees, and what has been learned from past experience. Sometimes the outfielder makes a mistake and comes in when he should go back, but that is the exception. The rule is that he catches the ball and sometimes by an extraordinary effort. The point is that the marvelous brain we possess allows us to do so many wondrous things that we take it for granted and don't really appreciate the amount of coordination, computing, and balance that goes into a successful catch, throw, or driving a car or doing the many things that we do day in and day out.

When we turn our attention to religious and spiritual concerns, the wonder is still present. When we pray silently or even aloud, how does the prayer get to God? When someone tells us that God wants him to do a certain thing or that God has spoken to him, how are we to understand such a matter? How does God communicate with us? It must be through our brain some way, but how are we able to distinguish between our own inner thoughts, feelings, wants, desires, and so on, God's speaking to us? How does the Holy Spirit guide and direct us? Undoubtedly through our brains some way, but we know not how. Often, of course, we are led by events when God either opens or closes doors for us so that, because of our prayer and devotions, we feel that we are led by the spirit of God. As a clergyman with some thirty years' experience, I often think that some religious people do not give proper consideration to the fact that the physical and the spiritual parts of human beings are inextricably bound together. And although such people may be correct when they proclaim that the Holy Spirit is guiding them, it may very well be that they are incorrect and that it is a part of their brain that is guiding them. It is my opinion that one should be more modest about such matters and say something such as "I feel that God is guiding me in such and such a direction," rather than being dogmatic about God's guidance. The apostle Paul was more careful at times about such matters as he was in quoting the prophet Isaiah: " 'For who has known the mind of the Lord so as to instruct him?' But we have the mind of Christ" (I Corinthians 2:16). So, since the human brain is so important in religious matters as well as in other concerns, it is incumbent upon all of us to know something about the brain and how it works.

It is difficult to overestimate the importance of the human brain in understanding ourselves and our actions. For example, there is the promise that research on the brain will yield the secrets of mental illnesses, give us a greater understanding of mankind, and free us from some of our misconceptions about ourselves and others.[1] Furthermore, the human brain is very special. It is not as large as the brain of some of the higher animals, such as the whale, but man's brain determines culture, allows consciousness, language, and memory, and, above all else, permits that marvelous human gift of speech that allows us to communicate with our fellow human beings. The brain is the basis of our imaginations, memories of the past, and planning for the future.[2] Restak puts the matter even more strongly as regards modern college curricula:

> Along with these theoretical concerns, it seems to me that psychobiology implies the adoption of certain practical measures. As things stand now, the typical graduate of most major universities knows next to nothing about the workings of his own brain. This is particularly ludicrous when you consider that all mental life is directly dependent on the integrity and optimal functioning of the brain. How absurd that a person, in order to understand the world, should study history, sociology, even psychology, yet posses no information at all about the regulator of all human activities.[3]

I can only add that everything we see, everything we hear, everything we read, every interpretation we make, every inference we make, every judgment we make, and every problem we solve or don't solve must be processed through our central nervous system—that is, through the brain and the nerves that feed into it and go out from it. So we can well say that the way we see and understand the world depends upon the human brain. The human brain will decide if we put missiles in Western Europe (which since I first wrote these lines is now taking place) and whether the Soviet Union responds by putting more missiles in Eastern Europe or elsewhere (which also has taken place since I first wrote these lines). In short, the human brain may very well decide whether or not to launch an all-out atomic attack that, if initiated, could spell the end of life on earth as we know it. Of course, there is always the possibility that through the prayers of millions of people God will intervene and prevent such a catastrophe from occurring. But these and other reasons certainly point out the supreme importance of the brain and our knowing how it works. And in this regard we are faced with a strange situation in which an organ seeks to understand itself. With our brains we are trying to understand how the brain works!

Now I should like to present some general information about the human brain. This is not meant by any means to be a complete description of how the brain works and of its parts. That information is found in other books, some of which I shall cite. Furthermore, I proceed on the assumption that the theory of evolution, while certainly not without gaps and far from complete, still makes sense and is a good working model. Many Christians will take issue, some vehemently, with me on this point, but I am prepared for such an outcome.

The relatively new science that deals with the human brain is called by various names. Some scientists call it psychobiology, some neurobiology, some neuroscience, and yet others seem to be content to refer to it as the study of the brain. The cells in the brain are called neurons and are the basic elements of the brain. The neurons communicate with each other by means of chemicals called neurotransmitters and by electrical impluses. Some of the neurotransmitters are: serotonin, acetylcholine, dopamine, and norepinephrine.[4] Neurons are connected with other cells by axons, which send out impulses to other neurons, and by dendrites, which conduct impulses to the neurons. Neurons are in communication with other neurons by means of synapses where the axons and dendrites do not touch each other, but there is a minuscule gap over which the impulses or chemicals travel. The neurotransmitters can either excite or inhibit the neuron from firing, and since each neuron acts as a small computer taking a sort of poll of the incoming impulses, it "decides" whether or not to fire. If it fires, the neuron will continue to do so as long as its threshold is exceeded.[5]

But neurons may be more complex than once thought. It has recently been shown that neurons can change from one neurotransmitter to another—for example, from norepinephrine to acetylcholine—and in the process of firing some neurons may change the ion from calcium to sodium.[6] Stevens gives us an idea of the complexity of just one neuron:

> The protein molecule (or complex of protein subunits) of the sodium pump has a molecular weight of about 275,000 daltons and measures roughly six by eight nanometers, or slightly more than the thickness of the cell membrane. Each sodium pump can harness the energy stored in the phosphate bond of adenosine triphosphate (ATP) to exchange three sodium ions on the inside of the cell for two potassium ions on the outside. Operating at the maximum rate, each

pump can transport across the membrane some 200 sodium ions and 130 potassium ions per second. The actual rate, however, is adjusted to meet the needs of the cell. Most neurons have between 100 and 200 sodium pumps per square micrometer of membrane surface, but in some parts of their surface the density is as much as 10 times higher. A typical neuron has perhaps a million sodium pumps with a capacity to move about 200 gradients of sodium and potassium ions that enable the neuron to propagate nerve impulses.[7]

In tracing the various impulses that go to the brain from hearing, touch, vision, and olfactory sensors, Nauta uses the analogy of a relay race, where a baton is passed from runner to runner and this banner arrives at the finish line in the same condition it started. In contrast, nerve impulses as they travel along the central nervous system are "recorded" as they pass each synapse, and the message arrives at its destination fundamentally changed from what it was when it started its course. Thus Nauta prefers to view the various synapses as processing stations.[8] Furthermore, the nerve impulses are all the same for the various senses, but the coding for each one is different. So the information relayed to the brain depends upon the stimulation of various clusters of receptors, which then send their coded message to the brain for interpretation.[9]

At the beginning of this chapter, I gave some illustrations of some rather complicated acts that happen with such regularity that we are apt to view them as being rather uncomplicated examples of behavior. But as we noted, there was a lot of activity going on in the central nervous system to allow such behavior to take place. But what makes our physical, mental, and, I am inclined to believe, also our spiritual activity possible is the large number of neurons and the manner in which they are connected. It is estimated that there are ten thousand million neurons in the human cerebal cortex. And according to Young, the really important thing about this vast number of neurons is that no two are exactly the same. Furthermore, each of the neurons adheres to one of the following conditions: (1) a small change going on in the outside world, (2) some part of a memory of a past external change, (3) some part of the instructions for an action to be carried out by the body.[10] And even though these cells may be different, they work together and influence one another because of the many interconnections mentioned earlier. The way the neurons are "wired" in the brain seems to be in a systematic manner rather than in some random fashion.

One of the reasons why we human beings are so adaptable is that the brain in general and particularly the cortex has at birth a large number of unwired gates, as Jastrow refers to them,[11] or uncommitted convolutions or neurons, as Penfield refers to them.[12] These unwired gates or uncommitted cortex allows the brain in man to develop in accordance to need and use. It may allow some people to become highly intelligent, some extremely agile, but all in all, it allows us to be extremely adaptable. It probably accounts for man's ability to adapt to weightlessness, to walk on the moon, and to be able to stand the pressures of exploring under the sea. Here I should like to digress and indulge in some speculation. We do not know what God has in mind for us human beings. It may be that mankind's penchant for exploration here on earth is partly a period of preparation for even greater exploration. It may be that God has in mind for us to really begin to explore the universe, and with the Voyager projects the process may have begun. And just recently there has been some speculation that the Russians are planning to land a man on Mars in the not too distant future. So is it possible that someday in the future—perhaps in the distant future—human beings may be born on spaceships heading toward another galaxy or for some particular star or planet? Is there intelligent life in outer space, or is it possible that in the entire universe planet earth is unique in its ability to spawn intelligent life? Of course, we don't know the answer to these questions, but if it should be that mankind does explore the universe, our extreme adaptability, due in part to our marvelous brain, will certainly be one of the salient reasons why we are able to do so. Meanwhile, we must be very careful not to destroy ourselves, our planet, and all the marvels, both animate and inanimate, that it contains. What a responsibility God has placed upon us as human beings; at the same time, what opportunity God has given us.

The brain functions so well because of the elaborate system of communication it maintains between the neurons. Stevens states that the communication takes place between axon and axon, dendrite and dendrite, and axon and cell body.[13] Furthermore, each neuron evaluates each signal reaching it. Some signals will say fire, some will say not to fire, and the neuron must make a decision. According to Taylor, the neurons use an averaging method that takes into account all the signals received, whether they are inhibitory or excitatory, and makes a decision whether or not to fire. Taylor refers to this method as parallel processing. So it appears that the neurons may use a sort of statistical method in arriving at a decision.[14] Jastrow compares the human brain to the modern computer by looking at the possible connections at gates in the two. The computer gate has at most four wires entering on one side of each gate. In contrast, every gate in the human brain is connected to as many as one hundred thousand other gates in other parts of the brain. So when a gate opens; it has had input from all the other gates with which it is connected.[15]

Restak believes that the brain is best understood in terms of three functioning units: alertness, information processing, and action. These three units must interact harmoniously for the brain to operate as it is meant to do. From Restak's point of view, it doesn't make any sense to ask where in the brain a certain activity is located, since brain activity is essentially an integrated process.[16] Nevertheless, there are certain areas, like the visual cortex, given chiefly to a single function. But the idea is that the brain functions as a unit. It is the processes going on in the brain that make it so complicated; this is due partly to the fact that the neurons affect each other so much. It may be that every part of the brain knows what the other parts are doing. So, as Taylor states, the paradox of the brain is that it is both localized and generalized; it analyzes and at the same time and with the same parts synthesizes.[17]

There is some question as to whether or not the human brain can be considered as some sort of hierarchy. There are what are sometimes referred to as higher and lower centers in the brain. For example, the cerebral cortex is the higher, the limbic system the next in line, and the brain stem the lower part of the brain. Young is not so sure that there is actually a hierarchy as such, although the brain is so structured that one can think of it in that way. Young points out that the spinal cord has the mechanisms for control of the body and limbs. The cerebellum has as one of its chief functions the responsibility to maintain balance and coordination of movements. The hypothalamus and its surrounding areas act as a sort of checkpoint to make sure that the brain does those things necessary to maintain the functions required for the maintenance of life.[18]

Bennett is more certain that the brain is organized in a hierarchical manner. He points out that the brain must be considered a dynamic organization distinguished by highly integrated structures. So he states that the various parts of the brain are organized into a hierarchy or "chain of command," with the levels affecting each other. The lowest levels are those that evolved earliest, having the function of controlling behavior for the basic needs of the individual and the species. The higher levels, such as the cortex, are responsible for making behavior more goal-oriented. Finally, the activity in any part of the brain can influence activity in any other part. So although the brain is made up of many diverse parts, it nevertheless operates as an organized unit.[19]

The brain has some unique characteristics, and I shall mention just a few at this time. The first of these characteristics is what Hunt calls "neural propensities or preprogrammed ways of dealing with experience." So that we can soon see and interpret various objects at a distance, we come to respond and cooperate with gravity and we recognize such concepts as time, space, and causality. And we soon adjust to our three-dimensional world.[20] Another characteristic of the brain worth noting here is its ability to release its own painkilling drugs. The reticular formation is a group of cells with small axons that are netlike in appearance. The reticular formation is located in the central core of the brain, and some of its cells release enkephalin, which is a neurotransmitter and serves to stop the sensation of pain. Enkephalin does this by stimulating the nerve cells that switch off the responses to pain and other stimuli.[21]

The last special characteristic of the brain that I wish to mention is the condition that all mental activity may result in brain change. Restak, applying this to psychotherapy, states that therapy may be creating changes in the patterned interaction of the brain cells. Restak quotes psychobiologist Gary Schwartz as saying,

"From this perspective verbal therapy is neurotherapy, even though we have no direct conscious awareness that this is actually the case."[22] This does indeed seem to be the case, as most people who have undergone psychotherapy know. It is not uncommon for people in therapy who are about to get some psychological insight to have a headache or an upset stomach, feel tired, feel out of sorts, be temporarily depressed, or not sleep as well as usual, the point here being that by releasing this insight into consciousness, the brain may undergo some slight change such as in the patterned interaction of the neurons as Restak suggests. Another possibility is that these insights may have to travel through the hypothalamus, for example, or pass through the corpus callosum—which connects the two hemispheres—say, on its way from the right to the left hemisphere. This process may be compared roughly to that of a blood clot that breaks off and travels to some organ where it makes some change in the operation of the organ. So too, when an insight "breaks off" from some part of the brain where it has been lodged perhaps for years and travels wherever it goes, it too makes some change in the brain, which is reflected in some of the symptoms mentioned above and possibly in some attitudes or behavior of the individual.

Now I wish to mention briefly some parts of the brain in order to show what part they play in our behavior and what their function is. The brain stem, in evolutionary terms, is an old and primitive part of the brain. It is concerned partly with sleep and arousal. The brain stem goes from the top of the spine to the center of the brain and may be thought of as the spinal cord entering the brain. The center of the brain stem contains some netlike fibers known as the reticular formation. The reticular formation has fibers that are both descending and ascending; that is, it sends signals both to the brain and away from the brain. The brain stem is concerned with the overall brain activity.[23]

The medulla, which is also located in the brain stem, plays an important part in regulating the internal environment of the individual, a condition called homeostasis. The medulla mediates regulation of heart rate, blood pressure, pupillary response, and other reflex actions.[24] A very important system in the brain is called the limbic system, named by Papez because parts of the brain involved form a ring. This system may have developed in an evolutionary stage after the development of the brain stem. The limbic system is made up of the following parts of the brain: the hypothalamus, the septal area, the cingulate gyrus, the hippocampus, the entorhinal cortex, most of the amygdala, and part of the thalamus.[25] The limbic system may be largely concerned with emotions, and some studies have linked the amygdala, a part of the limbic system, with the emotions of fear and anxiety and also suggested its role as a determinant of personality and behavior.[26] Restak reports that experts have linked the limbic system to hormones, drives, temperature control, reward and punishment, and memory formation. He adds that if you like simple ways to think about complex matters, you can think of the limbic system as the center for feeding, fleeing, fighting, and sexual behavior.[27] The limbic system is part of the old brain structure and is most often linked with emotions, some of which are not always useful in modern society.

One of the most important components of the limbic system is the hypothalamus. Bennett sees the hypothalamus as having two main functions. The first is in mediating behavior necessary for survival of the individual and ultimately the species. The hypothalamus makes sure that the individual gets enough to eat and drink and has input into sexual behavior. Second, the hypothalamus is the highest integration center for the autonomic nervous system. In this way, it controls much of our visceral function.[28] Calder sees the hypothalamus in much the same way and stresses its interrelatedness with other parts of the brain, such as the pituitary gland, through a network of nerves.[29] Leukel attributes five main functions to the hypothalamus: (1) regulation of the motor activity of the autonomic nervous system, (2) helping to maintain a constant internal environment in the individual, (3) regulation of endocrine output through connections to the pituitary gland, (4) controlling the basic drives, and (5) controlling emotional responses in rage and fear.[30]

Jastrow attributes these same functions to the hypothalamus, but also implicates it as the center of aggression, killing, and fight-or-flight responses. He cites studies where a rat that would normally kill a mouse placed in its cage will not do so after the hypothalamus is removed.[31] Young states that the hypothalamus is programmed to insist upon self-preserving activities and that it makes sure that we defend ourselves if attacked.[32] Taylor stresses the "pleasure center" role of the hypothalamus. Studies have been done in which a rat had a small electrode implanted in a particular part of the hypothalamus. When a small amount of electrical current was passed through the electrode to the hypothalamus, the rat got a pleasant feeling, and some rats have been observed pressing a small bar that allows electrical current to pass to the hypothalamus.[33] Some rats have pressed the bar so often and so long under such conditions that they neglected to eat; they would rather have their pleasure center stimulated than eat. So the hypothalamus, although a small part of the brain, about the size of a walnut in man, is very important part of the brain and appears to act as a sort of coordinating unit for many of the brain's actions among other functions mentioned above.

The cerebellum is another important part of the brain. It lies at the base of the brain on the back side. It is large compared to other brain structures. It is largely concerned with coordination of muscles to make the movements smooth and efficient. Once a skill has been learned, it is programmed in the cerebellum and kept for future use. Most of us have heard the expression, "Once you learn to ride a bicycle, you never forget." If we don't, it is because our cerebellum has filed away the skills and coordination necessary to perform the feat. So when we are driving or doing other activities requiring coordination, it is the cerebellum that is largely responsible for the fact that we can execute so many skills without being conscious of doing them or at least giving conscious thought to the activity. Jastrow states that of all the parts of the brain, the cerebellum comes closest to being an automatic computer.[34] Bennett notes the large size of the cerebellum and attributes this to two factors. First, the assumption of the upright posture and the discrete movements made possible by the evolution of the hands and especially of the opposable thumb.[35] We humans probably fail to realize what a wonderful development the opposable thumb is. Just imagine how many tasks and skills we couldn't do if we were not able to bring our thumbs and fingers together.

Perhaps the most distinctive part of the human brain is the cerebral cortex. Jastrow believes that the growth of the cerebral cortex began about 60 or 70 million years ago. He states that about that time, a band of forest mammals left the forest floor and went up into the trees, and this act, says Jastrow, was decisive event in the evolution of the human brain.[36] Young traces the truly human skulls of *Homo sapiens* from about forty thousand years ago.[37] So the growth of the cerebral cortex has been very gradual in mankind, but human intelligence evidently depends as much on how the parts of the brain are inwardly connected as it does on brain size.

What caused the human brain to grow so large and to have such complicated interconnected circuits compared to other forms of animal life? There are several reasons given for the growth of the human brain. We have already mentioned the upright posture as one good reason. The drive toward survival is certainly another one, since it required alertness, caution, fast reactions, skills, patience, resourcefulness, the need to find food, and the necessity to cooperate with at least some of one's fellow human beings. All of these must have put some kind of pressure on the brain to grow and to become more effective in all of the above requirements for survival. Another reason for the development of the size of the human brain was the chase. In order for early mankind to survive, they constantly had to be concerned with an adequate food supply. So to outwit animals that were faster, more cunning, and sometimes much stronger, early man had to find ways to bring down, trap, or catch his quarry. This requirement undoubtedly put more pressure on the human brain to grow.

The making and use of tools evidently was a big stimulus to the growth of the human brain. The evolution of language is often cited as one, if not the chief, reason for the growth of the human brain. One writer has suggested that war may be one of the chief stimuli to brain growth in the human being. When one stops and thinks about this proposal, one realizes that this assertion may not be too far off. We human beings have put a great deal of thought, planning, and technology into our development of destructive weapons. Just think of how much of our high-tech industry is geared toward the military. To make a weapon effective, many problems must be solved. In the long run, it may be that just the act of solving problems of all kinds forces the brain to grow.

Glasser believes that conflicts have a great deal to do with the growth of the human brain.[38] He particularly mentions the conflict between cooperation and competition that occurred in our remote ancestors. He believes that this conflict persists today and takes the form of a conflict between the desire to belong and share and the desire to compete and keep.

Jastrow mentions another interesting possibility for the growth of the cerebral cortex. The sense of smell is the only one of our senses that goes directly to the cerebral cortex without first being diverted to one of the other parts of the brain, such as the thalamus or hypothalamus. Our sense of smell has a direct connection with the cerebral cortex. Jastrow reminds us of how much our remote ancestors depended upon their sense of smell as they found their way about the forest floor. Since they may very well have traveled mostly at night, they had to rely upon their sense of smell to guide them and let them know just where they were and what each particular odor meant. Each odor conjured up some special picture or memory in their brains. All this places a special demand upon the size of the brain, states Jastrow.[39] We have similar sensations today when we perceive a particular odor. One odor may evoke pleasant memories, another odor may bring back sad memories, while yet another may elicit repugance. At any rate, for these and probably other reasons, the human brain, and here we are referring to the top part, the cerebral cortex, at various stages of our evolution grew at an accelerated pace. So it was the cerebral cortex that was the recipient of all the growth that such stimuli produced.

The brain is divided into two parts, usually called the right brain and the left brain or sometimes called the left hemisphere and the right hemisphere. The two halves are approximately symmetrical, although some writers have noted some small differences in the two hemispheres of the brain. The two hemispheres are connected for the most part by a band of fibers called the corpus callosum. It is via the corpus callosum that the two brains communicate with each other. There are over 2 million fibers in the corpus callosum; Restak estimates that if each fiber has an average firing of twenty impulses a second, the corpus callosum is carrying about 4 billion impulses a second.[40] This is an amazing amount of communication between the two brains, but it allows each of the two brains to know what the other is doing and thinking. Sometimes surgeons cut the corpus callosum to remove tumors deep in the brain. The corpus callosum is also cut at times to keep epileptic seizures from spreading from one hemisphere to the other, thus cutting down on the effects of the seizures. When the cerebral cortex grew, it needed to do so within the confined space of the skull. The way it grew in a confined area was to grow in folds called sulci (singular sulcus); the areas between the sulci are called gyri (singular gyrus). The cortex is divided into lobes, and each one is named. The lobes are called the frontal lobes, the temporal lobes, the parietal lobes, and the occipital lobes.[41] The right brain controls the left side of the body, and the left brain controls the right side of the body. One side of the brain or the other is dominant over the other side. The left brain is dominant for most right-handed people; while the reverse is true for most left-handed people; although for some left-handed people the left brain is dominant. I will have more to say about the matter of brain dominance in a later chapter concerned with conflict.

The cerebral cortex is organized vertically—that is, the

signals are sent up and down in columns something like the cells in a honeycomb.[42] The hemispheres can behave independently. This is shown in experiments with patients whose corpus callosums have been cut. Jaynes tells of one such experiment:

> This is one of the ways these commissurotomized patients are tested. The patient fixates on the center of a translucent screen; photographic slides of objects projected on the left side of the screen are then seen only by the right hemisphere and cannot be reported verbally, though the patient can use his left hand (controlled by the right hemisphere) to point to a matching picture or search out the object among others, even while insisting vocally that he did not see it. Such stimuli seen by the right nondominant hemisphere alone are there imprisoned, and cannot be "told" to the left hemisphere where the language areas are because the connections have been cut. The only way we know that the right hemisphere has this information at all is to ask the right hemisphere to use its left hand to point it out—which it can readily do.[43]

The two hemispheres have different functions. The left hemisphere contains the area that is responsible for language and speech and is the more analytic side of the brain. The right hemisphere has as part of its responsibility the expression and recognization of emotion.[44] The right brain is also better at recognizing faces. This must have been quite an asset in being able to distinguish between friend and foe in our early ancestors. And, of course, it is still a valuable aid to be able to look at someone's face and discern the emotions that he or she probably is feeling.

Patients who have suffered lesions in a particular part of the brain have been tested to see how these lesions affected their performance. It was found that damage to parts of the left brain affected speech, language, mathematics, sense of time, and verbal memory. In contrast, damage to the right brain affected performances in understanding visual and tactile mazes and perception of depth and movement and produced other disturbances.[45] The right hemisphere is called the "minor" brain, but it plays an important role in our lives. According to Restak, it is the right hemisphere that enables us to maintain a sense of who we are, to see things as a whole, to see the whole picture. It enables us to make important distinctions and grasp a situation in its entirety.[46] So damage or disease to the right hemisphere can interfere with many of the above mentioned functions.

Jaynes sees the right hemisphere as more involved in synthetic and spatial-constructive tasks in contrast to the left hemisphere's more analytic and verbal functions. The right hemisphere tends to see parts as having meaning only within a context, and as mentioned above, this hemisphere looks at the whole picture. Also on certain tests, the left hand, controlled by the right hemisphere, is very good at tasks such as categorizing shapes, sizes, and textures.[47] Jaynes also comments that women are less lateralized in brain function than men; that is, some brain fuctions in men that are assigned to one or the other of the hemisphere are more evenly divided between the two hemispheres in women.[48] From EEG recording Ornstein has concluded that the brain tends to turn off or not use as much of the unused side of the brain.[49] So, for example, if the individual is speaking, in right-handed persons, the left brain would be used and the right brain would be turned off to some degree.

Jaynes gives an interesting account of what can happen in an individual when the corpus callosum has been severed:

> If two different figures are flashed simultaneously to the right and left visual fields, as, for example, a "dollar sign," on the

left and a "question mark" on the right, and the subject is asked to draw what he saw, using the left hand out of sight under a screen, he draws the dollar sign. But asked what he has just drawn out of sight, he insists that it was the question mark. In other words, the one hemisphere does not know what the other hemisphere has been doing.

> Again, if the name of some object, like the word "eraser" is flashed to the left visual field, the subject is then able to search out an eraser from among a collection of objects behind a screen using only the left hand. If the subject is then asked what the item is behind the screen after it has been selected correctly, "he" in the left hemisphere cannot say what the dumb "he" of the right hemisphere is holding in his left hand. Similarly, the left hand can do this if the word "eraser" is spoken, but the talking hemisphere does not know when the left hand had found the object. This shows, of course, what I have said earlier, that both hemispheres understand language, but it has never been possible to find out the extent of language understanding in the right hemisphere previously.[50]

Finally, the prefrontal lobes will be mentioned as we end this brief and all too inadequate survey of the human brain. Some of what we know about the prefrontal lobes is due to an accident that occurred in the town of Cavendish, Vermont, on September 13, 1848. Phineas P. Gage was a gang foreman for the Rutland and Burlington Railroad, and in an accident a tamping iron was driven through his skull. The tamping iron entered just under Phineas Gage's left eye and emerged at the front of the skull. Phineas Gage lived, but with a noticeable change in personality. Previous to the accident, he was an industrious, reliable person. After recovery from the accident, Phineas Gage became fitful, used profanity, which he previously did not do, and seemed to care little for his fellow human beings. He was stubborn and made plans but did not carry them out. He changed so much that some of his friends said that he was no longer Gage.[51] Restak sees the prefrontal lobes as playing a decisive role in determination, deciding on action, and regulating behavior. He states that if one had to describe the prefrontal lobe in one word, it would be *purpose;* goal-directed behavior is lost after destruction of the prefrontal lobes.[52]

The brain has been compared to various things in the past. Earlier in this century, it was popular to compare the brain to a telephone switchboard. Now it is popular to compare the brain to a computer. Certainly a computer is a much more apt item to compare the brain to than the telephone switchboard, but there is no telling what the brain will be compared to as our technology improves and gives us more sophisticated products. Restak quotes Paul MacLean, a neurologist and psychobiologist, as saying that the human brain is really like an archaeological site with layers that developed in different times. The most recent development of the human brain, of course, is the cerebral cortex. The limbic system came earlier, and the earliest of all is probably the brain stem, with its various parts. Each of these three brains sees things in its own way and responds to the outside world in its own way. Each of the three brains amounts to interconnected computers, each having its own intelligence, subjectivity, and memory.[53]

Jastrow makes the point that the mammal's brain, and consequently the human brain, must have a tolerance for error. For example, every rustle of the trees or branches must not create an emergency; on the other hand, the brain must be able to detect a predator if one is lurking nearby. In short, the brain must not always be crying wolf when no wolf is near; on the other hand, it must be able to tell when a wolf is really near.[54] But the margin for error means that now and then the brain will be mistaken and not cry wolf when a wolf is really near. This margin for error in contrast to the computer's precision makes the brain much more flexible than the computer. Jastrow goes on to compare the two:

> The qualitative superiority of the brain over today's computer is even more striking than its compactness. Every cell, or gate, in the brain is directly connected to many other cells, in some cases to as many as 100,000. As a result, when we send a conscious impulse down to the recesses of the memory to summon forth a point of information, the cells in which this information is stored communicate on a subconscious level with thousands of other cells, and a wealth of associated images pours out at the conscious level of thought. The fruits of the subconscious activity are intuitive insight, flashes of perception and creative inspiration, all made possible by the countless connections among the cells of the human brain.
>
> The computer memory, in contrast, is like a set of pigeonholes stacked against a wall, with no thinking capacity in any pigeonhole, and no connections from one hole to another. Information can be placed in a pigeonhole or taken out of it, but there are no associations and thinking goes on elsewhere.[55]

Hubel assumes that the brain and computer are both machines and that they have some similarities as well as differences. But one of the most conspicious differences that he sees between brain and computer is in the elements, if synapses rather than neurons are taken to be elements in the brain. His guess is that the number of synapses in the human brain may number as many as 10^{14} (100 trillion), so he sees little possibility that computers will ever catch up.[56]

Finally, Taylor lists some advantages that the brain has over the computer. The first is that a computer has to be built by someone, it has to be programmed by someone, and it can be programmed only within rather narrow limits.

Furthermore, computers are not motivated, they have no curiosity, and they only tackle problems that they are instructed to tackle. The computer has no moral sense, no conscience, no generosity, no aesthetic feelings, no sense of wonder, and, worst of all, no sense of humor. They have no sense of values, attitudes, or preferences. On the other hand, the human brain constantly checks on itself. Humans can sleep on a problem and the unconscious can mull things over without any conscious effort on the part of the individual. Man has a memory stocked with experience and decisions that are checked against much of this past experience.[57]

So it seems that the human brain, which has been millions of years in the making, has many advantages over the computer. While it is true that in statistics, computing, and the like the computer can best the human brain, it is still the human brain that had produced the computer. So let us now turn our attention to something that is not as well defined as the human brain, but that is talked about as much, but much less understood than the human brain. I refer to that part of us that we call mind.

Chapter Two
THE ENIGMATIC MIND

When we turn our attention to the human mind, we find that we're in a different realm altogether. For the human brain is something that is made of flesh and blood; under the proper circumstances it can be seen and touched. But the mind is something altogether different. It cannot be seen or touched. In a way, the statement about the weather attributed to Mark Twain, "Everyone talks about the weather, but no one does anything about it," is to the point in talking about the mind. We may paraphrase the statement to read: "Everyone talks about the mind, but no one seems to know what it is." And this is just about the way the matter stands at present. Nowhere have I found this truth expressed better than in Julian Jaynes's book *The Origin of Consciousness in the Breakdown of the Bicameral Mind:*

> O, what a world of unseen visions and heard silences, this insubstantial country of the mind! What ineffable essences, these touchless rememberings and unshowable reveries! And the privacy of it all! A secret theater of speechless monologue and prevenient counsel, an invisible mansion of all moods, musings, and mysteries, and infinite resort of disappointments and discoveries. A whole kingdom where each of us reigns reclusively alone, questioning what we will, commanding what we can. A hidden hermitage where we may study out the troubled book of what we have done and yet may do. An introcosm that is more

myself than anything I can find in a mirror. This consciousness that is myself of selves, that is everything, and yet nothing at all—what is it?

And where did it come from?

And why?[1]

But when we ask the question, what is the human mind?, we find ourselves hard put to give an adequate answer. If we turn to *Webster's New Collegiate Dictionary*, we find that the following are the most relevant definitions for our purposes—"a: the elements or complex of elements in an individual that feels, perceives, thinks, wills, and especially reasons b: the conscious mental events and capabilities in an organism c: the organized conscious and unconscious adaptive mental activity of an organism." But to speak of the elements, events, and activity does not shed much light upon the question, what is the human mind? So we turn to some recent writers on the subject.

Taylor, in summing up his research on the mind, laments that scientists have not yet reached the stage where they can adaquately describe the mind and mental processes. He quotes several scientists as saying such things as "Over all, we are in a state of ignorance"; "Consciousness is just as unclear as ever"; and "Illusions are just as mysterious as they were in the nineteenth century"; and one scientist confesses "our ignorance of processes underlying the nervous system and the mind."[2] Another mystery of the mind is the mind-body problem. Is the mind something entirely different from the body, which is flesh and blood? Is the mind of the order of the soul or spirit? Either choice presents problems, as Hunt points out. If the mind is something like the soul or spirit, in what place does it store its ideas and conduct its processes? Why does it need a physical brain to carry on these processes? But on the other hand, if mind is only what brain does, how is mind aware of itself? How are we able to step back and to observe ourselves, and how is mind able to study itself? These questions are of much importance to all of us since they involve mortality or immortality.[3] For if mind is just what brain does, then death must be the end of the individual life. On the other hand, if mind is spirit and has some degree of independence from the brain, then immortality is a definite possibility.

Wilder Penfield, the Canadian neurosurgeon, as a scientist struggled for years trying to find a solution to the problem on the basis of brain action alone. But he finally decided that it made more sense and was more logical to adopt the hypothesis that the human being consists of two fundamental elements. He decided also that it would be impossible to explain the mind on the basis of neuronal action within the brain and that mind has a development all its own independent of the brain. His conclusion is, "I am forced to choose the proposition that our being is to be explained on the basis of two fundamental elements. This, to my mind, offers the greatest likelihood of leading us to the final understanding toward which so many stalwart scientists strive."[4]

On the other hand, Taylor notes many modern philosophers have moved to the position that the human being consists of one element, the physical. The major advantage is simplicity. For if the mind is held to be immaterial, three things need to be explained: the mind, the brain, and the interaction between the two. But if the mind is accounted for on a physical basis, then only the brain need be accounted for. Another difficulty encountered is the fact that if mind is immaterial, then there is the necessity of explaining how an immaterial object can act upon and influence a material object, such as when the mind exerts influence over the brain.[5] Personally, I fail to see the relevance of this objection. We readily accept the fact that immaterial things such as ideas, thoughts, desires, feelings, et cetera, can have a strong influence over parts of the body. To take an example, there is the Type A personality that is so often associated with heart problems. Many of the emotions held at least partly responsible for the heart problems, such as anger, lack of patience, a driving motivation, and a driving desire for success, are immaterial and exert considerable influence upon part of the body and brain. Also, it is readily admitted that anger and hostility can cause headaches or ulcers in some people. So it appears to me that the objection of how immaterial substances can influence physical materials is not really to the point.

Taylor sums up his concept of mind when he states: "The mind is not just the activity of the brain, in the sense that whatever activity is going on at any particular moment is all there is to it. Rather, the term mind refers to all possible activities of the brain, and it includes all the abilities and inclinations that are acquired over the course of development."[6]

Young is skeptical of the concept of mind and refers to it as "an entity called the mind." He states that the mind is perhaps that part of the brain's functioning organization of which we are conscious, and that at best "mind" is a vague concept.[7] In the latter regard, we certainly must agree with Young, but in the third part of this chapter I will attempt to show what we think mind is by the way we use the word in our English language. Taylor goes on to state that the mind is physical in that it is a reflection of activity within the brain and as such as much a natural phenomenon as the changing seasons. But it is also social and could not exist as it is now without the social pressures that have been applied down through the ages to help shape the human mind.[8]

Restak makes a distinction between mind and brain

that I find helpful. In making this distinction, he writes: "The mind and the brain are two aspects of a deeper, rather mysterious structure of human personality. When we apply psychological methods, we encounter 'mind'; when we opt for measuring neuronal activity with microelectrodes, we deal with the 'brain.' "[9] This distinction may be accurate, and at least theoretically it allows us to "think" about mind or brain. But in the long run the mind and brain may be so inextricably bound together that to separate them except on theoretical grounds is impossible. But it is the hopes and expectations of many scientists that in the future we will be able to know more precisely just what we are dealing with when we talk of mind and brain.

In the meantime, many of our mental functions are attributed to the mind. For example, Penfield says that the mind focuses attention, reasons, makes decisions, understands, and acts as if it has an energy all its own.[10] In reading the literature of brain and mind, one frequently gets the impression that brain is the servant of the mind or that the mind somehow uses the brain to carry out its purposes. Penfield also uses the analogy of the brain being the computer and the mind the programmer: "If decisions as to the target of conscious attention are made by the mind, then the mind it is that directs the programming of all the mechanisms within the brain. A man's mind, one might say, is the person. He walks about the world, depending always upon his private computer, which he programs continuously to suit his everchanging purposes and interest."[11] From this standpoint, we might say that brain is the servant and mind is the master, although I am certain that the relationship is not nearly that simple.

Can the mind be associated with any particular part of the brain? Some writers think so. Taylor, for example, believes that the mind is most closely associated with the newest part of the brain: "The newly evolved cerebrum is the part of the brain most closely associated with the activities of the mind; it is these activities which contribute most heavily to conscious experience. The cerebrum is known to be involved in perception, thought, memory, language, and many other functions of the human intellect."[12] On the other hand, Penfield locates the physical basis of the mind in the higher brain stem, since this is the part of the brain most closely associated with consciousness. Inactivation of this part of the brain can be effected by pressure, trauma, hemorrhage, and local epileptic discharge. Inactivation of the higher brain stem occurs naturally during sleep.[13]

Calder believes that there is nothing in current brain research that makes belief in the divine origins of the human mind incompatible with these findings. He quotes Sir John Eccles as stating, "I believe that there is a fundamental mystery in my personal existence, transcending the biological account of the development of my body and my brain. This belief, of course, is in keeping with the religious concept of the soul and with its special creation by God."[14]

So we find that the mystery of the mind remains. Is our nature that of two different elements, one physical and the other of a nonphysical nature, or are we of one element and only physical? Does the mind exist, or is it merely a reflection of what brain is doing? Perhaps science will someday be able to give us the answers to these and other questions concerning the true nature of man, his brain and mind. Until then, we must be content to live with the mystery of the mind.

Now we may ask ourselves how the concept of mind is used in the New Testament. One of the problems we face is trying to determine what "mind" meant to the New Testament writers is that most of them, with the exception of Luke and possibly the author of the Gospel according to John, wrote in Greek but had a Jewish background. And the Jews of the New Testament era did not have the same understanding of mind as we do today. So one of the problems is which Greek words translate into the English word *mind* and which translate into *thought, reason, will,* et cetera. This problem is expressed very well in *The Interpreter's Dictionary of the Bible:* "The translator's chief difficulty is that the Bible is the product of the Hebraic mind, which had no real interest in psychological analysis and no conception of the divisions of the human personality into separate organs or faculties, each governing some particular phase of man's psychic activity. Feeling, thinking, planning, and willing were all conceived to be functions of the entire personality, so that the conception of 'the mind' as the special seat or organ of reflective thinking as distinguished, e.g., from the heart as the seat of the emotions would have been, for the Hebrews, almost unintelligible."[15] So the words translated "mind" in the New Testament are best understood by context rather than by some other method. This is the method I shall use in trying to determine what the various New Testament writers meant by the various Greek words translated into English *mind*. For example, in the Gospels we find the English word *mind* in some verses, as in Mark 12:29: "Jesus answered, 'The first is, "Hear, O Isreal: The Lord our God the Lord is one; and you shall love the lord your God with all your heart, and with all your soul, and with all your mind, and with all your strength." ' " The Greek word translated *mind* here is *dianoia,* which has the idea of "the understanding" or "mind." In this context we may be safe in translating the word for *mind* as referring to one's intellect. Thus we are to love the Lord with all our emotions, with our soul or spirit, with our intellect, and with our physical strength. So, in this particular context, we may say that the use of the word in the

New Testament is very close to one way we understand the word *mind* in English today.

We find the word mind used again in Mark 5:15: "And they came to Jesus, and saw the demoniac sitting there, clothed and in his right mind, the man who had had the legion; and they were afraid." Here the word for *mind* is translated into the English phrase "in his right mind." But the meaning is clear. A man who had been "out of his mind" was now in "his right mind." A person who had been emotionally disturbed was now dressed properly and conducting himself in a "normal" manner. So, in this context, the word *mind* refers to one's emotional and mental stability, which again is very close to the way we use the word *mind* today on certain occasions.

The last example from the Gospels is from Luke 24:45: "Then he [Jesus] opened their minds to understand the scriptures,. . . " Here we may understand the word *mind* to mean that the two were able to "understand" or "comprehend"; it is almost as we say today when we sometimes receive an insight into something we had not previously understood. We may say, "Ah, ah," "Now I see," or something of the sort. So the phrase "opened their minds" has the meaning of Jesus making certain things clear to them and allowing them to comprehend and understand things that had not been clear to them before.

When we turn to the letters of Paul, we find that he uses most frequently the Greek word *vous* for mind. In Roman 7:22–23 we find this use of the word *mind*: "For I delight in the law of God, in my inmost self, but I see in my members another law at war with the law of my mind and making me captive to the law of sin which dwells in my members." Here the word *mind* is used in the sense of "intellect," "thinking," "inmost self," or, as we might almost say, "the real me." In Romans 12:2 we read: "Do not be conformed to this world but be transformed by the renewal of your mind, that you may prove what is the will of God, what is good acceptable and perfect." Here *mind* may mean reason or moral consciousness. The mind is something that may be "renewed" or perhaps be made "right again." In this sense, mind may be very close to what we call conscience today.

In I Corinthians 2:16, Paul wrote: "For who has known the mind of the Lord so as to instruct him? But we have the mind of Christ." In the first use of mind in the above statement, the meaning is close to asking, who knows the the intention or purposes of the Lord? The second use of the word *mind* in Paul's statement is better understood in another of Paul's letters. In Philippians 2:5–8, Paul wrote: "Have this mind among yourselves, which is yours in Christ Jesus, who though he was in the form of God, did not count equality with God a thing to be grasped, but emptied himself, taking the form of a servant, being born in the likeness of men. And being found in human form he humbled himself and became obedient unto death, even death on a cross." Here the mind "which is yours in Christ Jesus" refers to Jesus' obedience and humility and lack of desire to gain equality with God. In this case, mind is used to refer to an attitude or lack of personal ambition on the part of Jesus. *The Interpreter's Dictionary of the Bible* sums up Paul's use of the word *mind* in the following manner: "According to Bultmann's elaborate and subtle analysis, 'the mind' in Pauline thought means the whole of the self conceived as the subject of its thinking, feeling, and judging, whereas the Body is the same self regarded as the object of these activities. It is true, in any event, that for Paul, as for the rest of the Bible, 'mind' in some sense includes the whole man and can, in fact, often be taken as practically equivalent to 'character' (Rom. 1:28; 11:34; II Tim. 3:8)."[16]

In I Peter 1:13 we have still another use of the word

mind, where the author writes; "Therefore gird up your minds, be sober, set your hope fully upon the grace that is coming to you at the revelation of Jesus Christ." In this context, the use of *mind* stresses determination, courage, and faithfulness to their Christian cause. Finally, in Revelation 2:23 the author states: ". . . and all the churches shall know that I am he who searches mind and heart, and I will give to each of you as your works deserve." In this context, the word *mind* refers to the intellectual toughness, determination, and faith and also to the fact that Christ will discern the intention, tenacity, faith, and willingness of the churches to change. So the authors of the New Testament use the word *mind* in the way just mentioned and we can see that there is no one way the New Testament writers used the word *mind* or one meaning that they attached to it.

Now for the third and last part of this chapter I want to take some of our common English expressions and try to determine what we think *mind* is by the way we use the word in our everyday speech. The first expression is: I gave him a piece of my mind. In this use of the word *mind,* we mean that we told someone just how we felt about a certain matter. So mind is more allied with emotions or feelings in this particular usage. Another use of the word *mind* is: I have half a mind to do such and such. Here the emphasis is upon intention, desire, or the process of making up one's mind (which, by the way, is another use to which we English-speaking people put the word *mind* in which decision is the emphasis). This use of the word *mind* also implies indecision. The mind in this usage might be compared with a computer, which is weighing the evidence in order to make a decision. The phrase "always on my mind" has been used recently in popular songs and has to do with concentration, thinking about, mentally seeing someone's face, and the like. So in a sense this use of the word *mind* is very close to memory, so that we might say that "always on my mind" is tantamount to saying "always in my thoughts."

Another expression is "I had that in mind." Here *mind* is used as something where we may keep information stored, as a sort of reservoir in which memories, et cetera, are stored. So from this standpoint, we might say that the mind is a container. In the expression "he lost his mind," mind has to do with sanity, and to make sense, it cannot be taken literally. What *mind* refers to here is one's sanity, mental health, emotional stability, that sort of thing. So in our everyday speech, we often use *mind* to mean or at least to refer to sanity. In the expression "peace of mind" we are talking about feeling good about oneself, feeling comfortable with oneself and at peace with God and the world. Peace of mind implies that one has no significant problems or worries that cannot be overcome. "Tough-minded" is the next expression. Being tough-minded means that we are not easily hurt by others, that we may have a good deal of determination; it may mean that we are scientifically oriented. So being tough-minded may have to do with how one views the world or mean, as *Webster's New Collegiate Dictionary* defines the expression, "realistic or unsentimental in temper or habitual point of view." By extension, then, the expression "tough-minded" may refer to one's philosophy of life. In "I had a mind to", *mind* refers to intention or desire that was not carried out. That is, one may have considered several options and was inclined to do various things, but ruled out all but one. In one respect, we may say that this expression may mean that one was tempted to do such and such a thing. So in this expression *mind* refers to inclination, desire, or perhaps temptation.

Next is the expression "high-minded." *Webster's New*

Collegiate Dictionary defines this as: "marked by elevated principles and feelings." In this expression, *mind* has to do with one's principles, morals, or ethics and how one feels about them. In the expression, "keep that in mind," the mind is by implication compared to something that holds or contains things. The mind is retaining information that one may need to remember in making a decision or arriving at some conclusion. In the expression "it's all in your mind," *mind* may refer to various meanings. For example, it may mean it's all imagination, with no real substance to the matter. It may mean that the ache or pain is psychosomatic. And on occasion it may refer to the fact that someone believes you—or someone else—to be emotionally unstable. In the expression "mind over matter," we sometimes make the point that somehow, by concentration or determination, we can change, influence, or overcome certain physical obstacles. The implication is there also that the mental process has some ability to influence physical objects or forces.

The expression "in the back of my mind" again implies that the mind is something that can hold desires, information, inclinations, and the like. We sometimes use this expression when we want to convey the idea that we had not completely forgotten something or that we were keeping a certain alternative in mind. "Out of sight, out of mind" conveys the idea that if we don't see something, we forget about it; so in this expression, we come close again to equating mind with memory. In "she changed her mind," we again have the idea of mind as intention or perhaps even plans. In the expression "to read one's mind," *mind* is both subject and object, since it is one mind trying to "read" another mind. But this expression again really has to do with intention or thoughts. The mind as object is being "read" by another mind. The mind as object has to do the intention or thoughts, while the mind as subject has to do with the ability to discern intention or thoughts. The final expression is "feed the mind." We find this expression in such statements as "Great literature feeds the mind." Here the mind is something that needs nourishment, and if we remember the expression "voracious reader," we think of the mind as something that can devour or consume ideas, words, and the like.

Of course, this is not an exhaustive list of the English expressions using the word *mind,* but these seventeen expressions give us a good sampling of what we believe the mind is by the way we use it in our English language.

In this chapter we have examined the concept of mind on three different levels—first, as some of the scientists view mind; second, as some of the New Testament writers thought of mind; and finally as we use the word *mind* in our everyday speech in the English language. In the next chapter, we turn our attention to consciousness, which is every bit as much a puzzle as mind.

Chapter Three
CONSCIOUSNESS

When we turn to the concept of human consciousness, we find that we are on just as shaky ground as we were in dealing with the concept of mind. Winston Churchill is reported to have said of the Russians that they were a puzzle wrapped in an enigma wrapped in a conundrum. That description is also quite apt when speaking of human consciousness. For we don't seem to know just what it is, where it is, how it got there, when we as human beings became conscious, or if we were always conscious, and the answers to other questions that plague those who try to write or speak about consciousness. However, there are some very good opinions—some backed up by good evidence and reasoning—about these and other questions that various scientists and writers have put forth. So in this chapter we shall try to answer some of the above questions with some rather tentative answers and perhaps I should better say suggestions or possibilities. In any case, there are always many problems in trying to write or think about human consciousness. Part of the problem has been stated very well by Blakemore.

> For scientific, as well as philosophical, debate it is important to distinguish the state of consciousness from the so-called actions of the conscious mind, like choosing, whose existence as genuine operations Ryle and others would deny. It is this active aspect of consciousness that still gives scien-

> tists nightmares, philosophers headaches and theologians eternal joy. The question of choice is often the subject of acrimonious debate between reductionists and theists; so it is curious to hear arguments from both sides of the fence that the actions of man are entirely predetermined, either by the totally predictable and mechanical operations of the connections in his brain or by the predistination written for us all in the diary of an equally mechanical God. This attitude, whatever the motivation behind it, is an insult, either to the subtlety of the brain or to the mentality of God.[1]

If we include the problem of freedom of choice or free will, as Blakemore has done, the problem of consciousness becomes even more ponderous. To pursue that aspect of consciousness would take us too far astray from our main topic of consciousness itself, so suffice it to state that there is plenty of evidence, both from the standpoint of how our brains are connected and from the biblical point of view, to make a good case for either freedom of choice or determinism.

When it comes to asking just what consciousness is, Taylor gives us a straightforward answer: "We know from experience what consciousness is: it is the continuing, flowing awareness of ourselves and the world around us that lies at the heart of our existence as human beings."[2] But he goes on to state that beyond that, we know very little about consciousness, since we cannot see it, hear or touch it, or know it through any of our other senses. We cannot compare our consciousness with that of others except insofar as we agree on what we see, hear, smell, and the like. Some writers go so far as to say that since we experience consciousness in ourselves, we infer that others have consciousness, too. Earlier in this chapter, I quoted Blakemore as stating that consciousness still gives scientists nightmares, philosophers headaches, and theologians eternal joy. He elaborates on the scientists' dilemma when he states:

> Most scientists are embarrassed when they cannot explain events by the forces and laws that they already understand. Those who study the brain usually shuffle their feet uncomfortably and quickly change the subject when the discussion turns to that one brain function that we all know so intimately—consciousness itself. If *Things* cannot be hostile, how can the bits and pieces of molecular hardware that make up the brain have emotions and thoughts? The brain researcher of today is almost as impotent to evaluate consciousness as a computer is to judge beauty or put a price on a Rembrandt portrait. But it does not follow that beauty is more than the sum of a number of definable features, nor that the Rembrandt is more than all its individual brush strokes. The problem is to define all the brush strokes of the brain before we reject an opinion like that of W. Somerset Maugham, who said: "The highest activities of consciousness have their origins in physical occurrences of the brain just as the loveliest melodies are not too sublime to be expressed by notes."[3]

This idea of consciousness having its origin in the physical brain is not incompatible with modern research on the matter, as we shall see directly.

One of the first things to be noted about consciousness is that it is concerned with our own inner world: our own feelings, impressions, how we see the outer world and our own personal evaluations of the physical world. For we must remember that everything that we see, hear, feel, touch, or smell goes through our own personal nervous system. So consciousness is a very personal thing for each of us. Taylor points out that impressions about the external world may be included in our concept of consciousness, but that it is actually our own mental activity of perceiving that is reflected in consciousness.[4] It is generally agreed that consciousness is concerned with process, not structure.

Gordon Rattray Taylor believes that consciousness is not an all-or-nothing thing. He relates an account of the recovery of a soldier who had received a severe head wound in Normandy during World War II. The account was gotten from Hugh Cairns, a brain surgeon. The soldier was wounded on July 16, 1944, and he was unconscious. On July 23, when called by name he would open his eyes; on July 26, he was fairly alert, but did not speak well. By August 5 he could hold short conversations, and so the progress went.[5] Taylor could be right in his all-or-nothing theory; on the other hand, the slow recovery could merely be a reflection of nature's slow manner of healing. When we are seriously ill or wounded, we are naturally in a hurry to get well, but nature is in no such hurry. This may indeed be a reflection of the way God works in his universe, for if the world is some four and half billion years old and the universe is twenty billion years old, as Jastrow believes, then we—some of us, at least—must say that God is certainly in no hurry.[6]

Ornstein believes that consciousness is personal construction; that is, that each of us constructs or shapes or makes his own personal consciousness. This construct was made orginally for survival purposes, especially in order to read or understand the meaning of discrete objects, sensitivity to forces that may pose a threat and to separate oneself from others.[7]

Jaynes gives us another reason why consciousness may have been valuable in the struggle for survival:

> Overrun by some invader, and seeing his wife raped, a man who obeyed his voices would, of course, immediately strike

> out, and thus probably be killed. But if a man could be one thing on the inside and another thing on the outside, could harbor his hatred and revenge behind a mask of acceptance of the inevitable, such a man could survive. Or, in the more usual situation of being commanded by strangers, perhaps in a strange language, the person who could obey superficially and have "within him" another self with "thoughts" contrary to his disloyal actions, who could loathe the man he smiled at, would be much more successful in perpetuating himself and his family in the new millennium.[8]

Now, of course, many of us are experts at hiding or disguising our real feelings. We may do this when being dressed down by our boss or any other superior or around a man or woman whom we are fond of but to whom we are not yet ready to show our real feelings. But this capacity to disguise feelings is sort of a mixed blessing, for down through the ages we have learned to hide our true feelings even from ourselves and nature takes a dim view of dishonesty even when the person we deceive is ourselves—in some cases, we might better say, especially when it is ourselves that we deceive.

Blakemore makes the point that evolution has given us internal awareness, which must have survival value. He notes one obvious advantage that consciousness gives us—namely, the ability to make predictions about the behavior of other people.[9] The study of body language is concerned with this very thing. We would like to know what others are thinking or if they are favorably disposed toward us or not. Some poker players are very good at this and can tell pretty well if the other players have a good hand or if they are bluffing. But to put the matter on another and higher level, imagine the advantage one would have in an important business deal if he or she were able to tell if the other company's representatives were ready to sell at a lower price than previously mentioned.

Or even more urgent would be a summit meeting between the president of the United States and the chairman of the USSR. If one of them were able to discern the true feelings of the other, the country would have a definite advantage over the other in the negotiations. Part of such discernment may be innate with some people, but it can also be learned, and it surely will be a part of our education in the not too distant future—at least for some people and for some vocations.

Hunt gives us a cognitive view of consciousness:

> Cognitive science does offer at least a rudimentary explanation of consciousness: it is thought to be the product of our internalizing the real world in our minds in symbolic form. We perceive not only the real world but also our own mental representation of it; the experience of the difference between the two results in self-awareness. We recognize that there is not only a real world but a simulacrum of it within us; therefore there must be an *us: Cogito ergo sum,* yet again. The thought that we have thoughts is the crucial one that becomes consciousness; it is what Douglas Hofstadter, in *Godel, Escher, Bach,* calls a "strange loop" of the mind. On interaction between different levels, a self-reinforcing resonance. "The self," he says, "comes into being at the moment it has the power to reflect itself." We contemplate our thoughts but the awareness of doing so is itself a thought, and the foundation of consciousness.[10]

So from this point of view consciousness is partly the product of our internalizing the outside world in our minds and partly the awareness of our own thoughts. We can reflect and evaluate our own thoughts, and this is a sort of awareness of consciousness.

In writing about the various views to account for consciousness, Jaynes presents the metaphysical point of view and notes the vast difference in the culture and lives of human beings compared to the animals. He writes:

> The chasm is awesome. The emotional lives of men and of other mammals are indeed marvelously similar. But to focus upon the similarity unduly is to forget that such a chasm exists at all. The intellectual life of man, his culture and history and religion and science, is different from anything else we know of in the universe. That is fact. It is as if all life evolved to a certain point and then in ourselves turned at a right angle and simply exploded in a different direction.
>
> The appreciation of this discontinuity between the apes and speaking civilized ethical intellectual men has led many scientists back to a metaphysical view. The interiority of consciousness just could not in any sense be evolved by natural selection out of mere assemblages of molecules and cells. There has to be more to human evolution than mere matter, chance, and survival. Something must be added from outside this closed system to account for something so different as consciousness.[11]

This is the very problem that those who take a purely evolutionary point of view toward life on earth run into: all forms of life developing by chance, not to mention consciousness itself. To take one example, what would be the probability—or the improbability, if you will—of a world where plants give off oxygen and use carbon dioxide and animals use oxygen and give off carbon dioxide coming into being by mere chance? If we take enough of these examples from our world, we soon run into a huge amount of statistical material. Even the best of our modern computers might be severely overloaded by trying to figure such probabilities. To approach the matter facetiously, to believe that our world—not to mention the universe—could have been brought into being by chance takes more faith than most religious people can muster. The same thing might very well be said about consciousness itself.

Finally, we may ask the question, is consciousness associated with some particular part of the brain? Most writers and scientists that I have read think so. A few place the site of consciousness in the cortex. Most, however, place the site of consciousness in the higher brain stem. On this point Penfield has written: "Gradually it became quite clear in neuro-surgical experience that even large removals of the cerebral cortex could be carried out without abolishing consciousness. On the other hand, injury or interference with function in the higher brain-stem, even in small areas, would abolish consciousness completely."[12] Gordon Rattray Taylor agrees that the site of consciousness is in the brain stem, and he locates it more precisely in the pons of the brain stem. He relates that one fourteen-year-old boy fell and ruptured an artery that caused a blood clot that was pressing upon the brain stem and rendered the boy unconscious. When the blood clot was removed, the boy regained consciousness "and became embarrassingly lively."[13]

So although consciousness essentially remains a mystery, we have been able to say some things about what some writers and scientists believe it to be and where it resides in the brain. In speaking of consciousness, we would be remiss if we did not say something about the unconscious, which is just as hard to pin down and quite controversial. Let us then turn to that aspect of our mental activity.

As Gordon Rattray Taylor points out, we use the word *unconscious* in different ways. Sometimes we mean that the person has been knocked unconscious, and sometimes we mean that there are thoughts, ideas, desires, conflicts, and the like of which we are unaware. They are somewhere in our minds or brains, but we do not know they are there, as long as they remain unconscious. Once they become conscious—that is, we become aware of the ideas, thoughts, desires, fears, et cetera—then they are no longer unconscious. I once had occasion to know firsthand what the first definition of the word *unconscious* means. When in high school, I was practicing football one afternoon and I was carrying the ball. I tried to turn sharply and slipped, hitting my head on one of the other player's knee. I was knocked out, as we say, and when I came to, I was standing on the sideline talking to the school principal. It was a strange feeling to "wake up" and find myself on the sideline talking to the principal; how long I was out I don't know, but it couldn't have been very long. I was, as the saying goes, "out on my feet."

But that is not the kind of unconscious we are talking about in relation to consciousness. We are talking about that unconscious that Freud popularized near the end of the last century and the early part of the present century in such writings as *The Interpretation of Dreams* and *The Psychopathology of Everyday Life*. In this use of the word *unconscious*, we usually mean that our minds have hidden something from us. It may be a desire, fear, mental conflict, or the like. Again from my personal experience I am able to provide an example of this kind of use of the word *unconscious*. During World War II, I was on a train in early 1943 going from Oran to Algiers. We stopped at a water station to get water for the train. We were low on drinking water in our boxcar, and it must have been my turn to get the drinking water. Anyway, I got the water for our boxcar. At the same time, there was a train stopped at the same place going the other direction, west, loaded with German prisoners. They had their cups extended through the bars and were calling out, "Vasser, Vasser." I was sorry for them and carried several buckets of water to them until I thought our train was about ready to pull out, at which time I got back in our boxcar. I had to run a short distance to get each bucket of water and take it to the thirsty Germans. What I did not realize at the time and for some forty years was that I was afraid that our train might start up and leave me while I was making one of my trips for water. We were at a small station somewhere between Oran and Algiers; just where I have no idea. So the prospect of being left there alone was a rather frightening one, but the fear was somewhere in the uncounscious part of my mind or brain until just recently. This is the kind of unconscious that I am speaking of at the present time.

On the matter of the unconscious, Gordon Rattray Taylor has written:

> We have established two facts about the mind and consciousness, both of great significance, yet seldom recognized. First, that we must make a real distinction between conscious awareness and unconscious awareness. It is time we conceded openly that the latter exists. Most of our analysis of situations and our appraisal of the environment goes on at the unconscious level. Often we see something but fail "to take it in," as we say. We have hardly any awareness of perceptual processes (such as interpreting size constancy) or of memory processes (such as how we find a word which is on the tip of our tongue). And of course we have little direct conscious awareness of our autonomic responses. (Your finger swells when you hear a loud sound. Did you know?) Much information is held unconsciously, and it seems to be a function of consciousness to bring into conjunction things which the unconscious has failed to relate. Hence it seems that feeling and remembering are not in themselves manifestations of consciousness in the narrower sense. Most discussions of consciousness ignore such facts.[14]

Restak states that each hemisphere may have somewhat independent operations when it comes to interpreting certain material. He gives the example of a mother and her young daughter, in which the mother says to the daughter, "I am doing this because I love you, dear," while the face and voice tone say, *I hate you and will destroy you.* Here the left hemisphere would process the words "I love you" and make the daughter feel assured. But the right hemisphere is processing the facial expression and voice tone, which make the girl feel anxious and uncertain. So the daughter is getting mixed messages from the mother. The message the right hemisphere gets, *I hate and I will destroy you,* is the one that will be processed in the unconscious.[15] Elaborating on the right hemisphere as the possible site of the unconscious, Restak writes:

> "Observations like the foregoing lead us to favor the view that in the minor hemisphere we deal with a second conscious entity that is characteristically human and runs in parallel with the more dominant stream of consciousness in the major hemisphere," says Sperry. Could this "second conscious entity" in the right hemisphere be none other than Freud's unconscious?
>
> Dr. David Galin, of the Langley Porter Neuropsychiatric Institute in San Fransisco, thinks the right hemisphere's performance is strikingly similar to the operation of "unconscious" process. For one thing, both the right hemisphere and the unconscious deal with images which cannot ordinarily be verbalized. Second, their functioning depends less on logical analysis than on the perception of total pictures, which psychologists refer to as Gestalt.[16]

I am inclined to agree that the right hemisphere might be the site of the unconscious. But if it is, then we have to consider what might happen when material that has been unconscious becomes conscious. For one thing, such thoughts, ideas, feelings, fears, and the like probably pass through the corpus callosum to the left hemisphere. But since material passing from the unconscious to the conscious level often causes physiological disturbances such as headaches, upset stomaches, fatigue, insomnia, heart disturbances, and the like, such material on its way to the right hemisphere might also be routed through the hypothalamus, pons, brain stem, and perhaps other parts of the brain associated with bodily functions.*

Restak gives further evidence that the right hemisphere might be the site of the unconscious. He cites studies done by Gary Schwartz in which subjects were first asked factual questions and later emotional questions. For example, if asked how many s's in the word *Mississippi*, most subjects turned their eyes to the right, a sign of left-hemisphere activation. But when asked such questions as, Do you think your mother-in-law is an interfering woman? the eye movements were usually to the left, indicating right-hemisphere activity in response to an emotionally laden question.[17] Taylor has a rather novel answer to the perplexing question, How does some material become unconscious? His answer is:

> In a variant of this experiment [by Ann Treisman at Oxford], lists of words were read to each ear, so arranged that if you took a word from each ear alternately, they would combine to make a meaningful sentence. It turned out that the subject immediately spotted this and could report the sentence. So we are forced to the conclusion, and it is a very radical one, that much, if not all, of the input to our brains is monitered at the unconscious level, and filtered, before the interesting bits are supplied to consciousness.
>
> Now the only mechanism capable of making such a fine discrimination is the cortex itself. So we are driven to

*I have based this information upon personal experience. I have found that the physiological distrubances precede the psychological insights. That is, one would first experience a headache, upset stomach, or the like and then later become aware of the material that previously had been unconscious.

> conclude that the data reaches the cortex without conscious awareness and is then examined; when something of interest is found, a message must go down to the arousal mechanism in the brain-stem, which then fires back a "wake-up" instruction to the cortex to pay heed. This is quite heterdox, but no alternative has been offered. In fact, everyone has kept very quiet on the subject.[18]

Taylor may be correct on this solution to the problem, but I am inclined to believe that it may just as well be that there is something like Freud's "censor" that operates in the unconscious to either let certain information pass through to the conscious or keep material from reaching consciousness. Now in either case, the decision must be made instantaneously; and since there is no time for lengthy decision making, some rather trivial matter often gets retained in the unconscious. It must be something like a driver who has an emergency and cannot take time to think; he must act immediately or it may cost him his life. Now that unconscious, if that is actually where the decision is made to retain or pass information, may take some material as being in some way threatening when in fact it may not be so. At any rate, for some reason, and this decision is probably based more on feeling than thought, the material is deemed to be unacceptable to be admitted to consciousness.

If this is actually the case, then this must be one of the survival techniques with which we have been endowed. In this regard, I am reminded of one of Jesus' statements to the disciples: "I have yet many things to say to you, but you cannot bear them now" (John 16:12). Perhaps a decision is made somewhere in our brains from time to time that there are just certain things that we cannot bear at certain times or stages in our lives. This mechanism may be a very old one and go back to the time when we as human beings were just beginning to develop consciousness, and at that state of our development, it was undoubtedly true that there were many things that were too much for us to bear. And this is still true to a large extent for us today, for it is in the realm of the emotions and feelings that we have so much difficulty even today. So I conclude, then, that keeping certain material in the unconscious was originally—and to a large degree still is today—a protective measure with which we have been endowed. Now let us turn our attention to the Adam and Eve account in the Old Testament and see if that story has anything to say to us about consciousness.

First I shall briefly recapitulate the story of Adam and Eve as told in Genesis 2:15–25; 3:24. God created Adam out of dust and placed him in the Garden of Eden. Then God created all the other living creatures. Next, so man would not be alone, God created Eve out of one of Adam's ribs. Adam and Eve were allowed to eat some of the fruit from the trees, except they were not to eat from the tree of the knowledge of good and evil. However, the serpent persuaded Eve to eat some of the fruit from the tree of good and evil and then Eve gave some of the fruit to Adam and he ate also. Now before they ate from this tree, they were naked and were not ashamed. But after eating from the tree of good and evil, they realized they were naked and used fig leaves to cover themselves. When God saw that they knew they were naked, he asked if they had eaten from the tree of the knowledge of good and evil. Adam blamed Eve for his eating the fruit and indirectly God himself, since God made Eve for him. Eve then passed the blame on to the serpent. Adam and Eve were then driven out of the Garden of Eden and were not able to return, since the entrance was guarded by the cherubim with a flaming sword.

This account is usually called the fall of Man and is cited to account for man's sinful condition. This account of Adam and Eve also tells why women must bear children in pain, why man must earn his living by the sweat of his brow, why the ground has thorns and thistles and is hard to work, and why men and women no longer live in the Garden of Eden. But our main concern here is with the account of Adam and Eve in the Garden of Eden and what it tells us about human consciousness.

Before Adam and Eve ate from the forbidden tree of the knowledge of good and evil, they were not conscious of being naked and lived in an innocent, childlike state. After they ate from the tree of the knowledge of good and evil, however, they realized that they were naked and covered themselves. This can very well be interpreted as the difference between mankind before and after consciousness developed. It is to be noted that one of the chief results of becoming conscious was they they had the capacity to experience shame and guilt. But it seems to me that the very act of becoming conscious of oneself, of others, of one's existence, and of gaining knowledge of good and evil—all of this new consciousness and awareness exact a price from us as human beings. In short, becoming conscious means that one must pay a price, or perhaps we may say that there is a penalty for becoming conscious. I mentioned earlier that most people who have undergone psychotherapy know that gaining insight into oneself exacts a toll. We have to pay a certain price for becoming aware of our inner motives, hidden thoughts, desires, and conflicts and for tapping into the uncounscious. The penalty is often in the form of some physical distress, such as headaches, upset stomaches, tiredness, perhaps irregular heartbeat, or maybe arthritis of a temporary nature, and most certainly anxiety. But whatever the nature of the upset, it is some sort of penalty we have to pay for gaining added awareness or consciousness of ourselves. It is as though nature has decreed that we should know so much about ourselves and no more unless we are willing to pay a price. Perhaps this is just one more reason why there is an unconscious: It is a way of allowing most of humanity to avoid coming to grips with much of the forbidden knowledge or awareness of oneself. To consider the matter in that way adds more meaning to the saying: Ignorance is bliss. There are probably many things about ourselves that fit into that category, and it is much easier and less taxing to dump much of our mental excess baggage into the unconscious and never become aware or conscious of it at all. It is often more convenient to do so, unless, of course, one is required to drag it all out at some later date.

Perhaps in our investigation of consciousness, we have hit upon one of the basic reasons why we human beings have such a tremendous capacity to feel guilty. It is almost as though we had gone beyond certain set limits. But there have always been daring people who were willing to take risks, and I feel certain that this will be the case in the future, so that we will continue to learn more about ourselves and become conscious of more and more both in the inner world and nature itself.

Chapter Four
HUMAN INTELLIGENCE

When speaking of brain, mind, consciousness, and the like, we cannot ignore the subject of intelligence, for intelligence has so much to do with what we humans have become and what we might become in the future. If we take seriously one of the scientific designations of mankind, *Homo sapiens*, it is not hard to know what we think of ourselves. We are the wise ones. We are the intelligent ones. We are *Homo sapiens*. And with all due humility, I think that in some degree we merit that designation. If we look at all of mankind's positive achievements in the areas of science, engineering, agriculture, art, music, literature, religion, inventions of various kinds, medicine, and social achievements, we must say that the compliment we pay ourselves is not too much in excess. If, on the other hand, we look at some of our less noble activities, such as our making better and better weapons with which to kill ourselves, our inhumanity to some of our fellow human beings, our slums, our decaying inner cities, our indiscriminate use of chemicals on the land (which get into our drinking water, the oceans, the lakes, and streams), our pollution of the air, and some other acts that may, at least in the eyes of some, border on stupidity, then we may indeed seriously doubt that we have much intelligence at all and consider that we are far from being wise. Be that as it may, it is true that we do have what is called intelligence; perhaps the difficulty lies in the fact that we do not always act intelligently. I will have more to say about that matter in the next chapter.

Intelligence is somewhat easier to define than mind, or consciousness; still there is no unanimity concerning just what it is. If we turn to *Webster's New Collegiate Dictionary,* we find two germane definitions: "(1): the ability to learn or understand or to deal with new or trying situations; reason; also the skilled use of reason. (2): the ability to apply knowledge to manipulate one's environment or to think abstractly as measured by objective criteria (as tests)." Wechsler has a definition that I rather like: "Intelligence, operationally defined, is the aggregate or global capacity of the individual to act purposefully, to think rationally and to deal effectively with his environment."[1] Wechsler goes on to say:

> Although intelligence is not a mere sum of intellectual abilities, the only way we can evaluate it quantitatively is by the measurement of the various aspects of these abilities. There is no contradiction here unless we insist upon the identity of general intelligence and intellectual ability. We do not, for example, identify electricity with our modes of measuring it. Our measurements of electricity consist of quantitative records of its chemical, thermal and magnetic effects. But these effects are not identical with the "stuff" which produced them. We do not know what the ultimate nature of the "stuff" is which constitutes intelligence but as in the case of electricity, we know it by the "things" it enables us to do—such as making appropriate associations between events, drawing correct inferences from propositions, understanding the meaning of words, solving mathematical problems or building bridges. These are the effects of intelligence in the same sense that chemical dissociation, heat, and magnetic fields are the effects of electricity; but psychologists perfer to use the term "mental products." We know intelligence by what it enables us to do.[2]

This is definition more by analogy and description, but I think Wechsler's statement: "We know intelligence by what it enables us to do," is quite appropriate and gives us sort of a common ground for what we shall be discussing. I assume that Wechsler means that we know intelligence by the *constructive* things it enables us to do, and that is the way I interpret the statement. For although it takes a great deal of skill and intelligence to make our modern, sophisticated instruments of destruction, I don't think that it is necessarily behaving intelligently.

Some scientists believe that human intelligence developed over a period of a million years or more. Since the brain is about as large as it can get in the human skull, many scientists belive that human biological evolution may be over. But this may not be so in the case of the brain, since there is always the possibility of various components in the brain, such as the hemispheres, the hypothalamus, et cetera, making further refinements or changes for the better. Also there is the probability that the connections within the brain will continue to develop and make refinements. And since human intelligence may depend almost as much upon inner connections within the brain as brain size itself, intelligence could still be in an evolutionary stage. And here I am reminded of a sign I once saw that stated: Be patient, God is not through with me yet. In the same way, I do not belive that God had done all he wants or can do with us *Homo sapiens* yet; we may yet become even more sapient. Perhaps we may in time even become truly wise. But Jastrow is among those scientists who believe that human evolution may be about over. But he does not necessarily believe that the evolution of intelligence is over, and it is likely that it is not over. He puts the matter this way: "Apparently, among all traits of a living organism, none has greater survival value than the flexible, innovative response to changing conditions that we call intelligence. It seems unlikely that this trend in evolution, which has persisted for more than 100 million years, should suddenly stop at the particular level of mental achievement that we call 'human.' If the past is any guide to the future, mankind is destined to have a still more intelligent successor."[3]

If the world is some 4 billion years old, why did it take so long to develop human intelligence? Sagan has developed a cosmic calendar in which intelligence developed late in December, and in trying to account for the late occurrence of intelligence, he lists four possibilities: (1) never before was there a brain so massive; (2) never before was there a brain with so large a ratio of brain to body mass; (3) never before was there a brain with certain functional units (large frontal and temporal lobes, for example); (4) never before was there a brain with so many neural connections or synapses.[4]

Some of our ancestors are thought to have spent considerable time in the trees. Sagan writes about how dangerous it is for human beings to fall from much height and the fact that our ancestors in the trees had to be very careful about getting around in the trees. They had to have good vision and timing and be able to consider whether or not a limb would support their weight; in a word, a mistake could have been fatal. Sagan states: "But each of these skills required significant advances in the evolution of the brains and particularly the neocortices of our ancestors. Human intelligence is fundamentally indebted to the millions of years our ancestors spent in the trees."[5] Those of us who had the opportunity to play in trees when we were growing up know that one has to be very cautious in moving around in trees and especially in trying to go from one tree to another. Perhaps our love of playing in trees, not to mention tree houses as children, goes back to the vast time our ancestors spent as tree dwellers. In this regard, one may wonder if the human race is being prepared for further advances by driving cars and flying in planes, where a mistake can be and often is fatal. The training is much the same as that of our ancestors in the trees in that driving and flying require skills that can get one safely where he wants to go. Timing, good vision, coordination, and good judgment are much the same skills that were required to navigate in the trees. Of couse, it took much more strength to hang by one's arms in the trees than it does to sit and drive a car, and in that regard there is a distinct difference in the two skills.

In trying to get at the basis of intelligence, Sagan has suggested that the ratio of the size of the brain to total body mass is a good indicator. He writes: "The difference in brain mass between the sexes is of interest in precisely this context, because women are systematically smaller in size and have a lower body mass than men. With less body to control, might not a smaller brain be adequate? This suggests that a better measure of intelligence than the absolute value of the mass of a brain is the ratio of the mass of the brain to the total mass of the organism."[6] Also partly responsible for human intelligence, Sagan believes, are the microcircuits that require a very small amount of electricity to function, about one-hundreth of the voltage required to activate the ordinary neurons. This enables the microcircuits to make subtler and finer responses than the regular neurons are capable of doing.[7] But if we try to locate intelligence in some particular part of the brain, once again we are disappointed. As Wechsler has pointed out, studies indicate that no single part of the brain can be said to be the locus of intelligence.[8]

Sagan believes that the difference in functions of the

two hemispheres came about through long years of evolution. That is, the gain in manual dexterity also advanced the speaking ability and the other way around as well. Each time a gain was made in one area, it had an enhancing effect on some other skill. So the use of tools, the upright posture, speech, and thinking all worked together to help specialize the two hemispheres. The result is that for most of us, the left hemisphere is the dominant hemisphere and is the analytical part of our brain while the right hemisphere has developed a more intuitive mode of thinking.[9] Furthermore, our ancestors must have used the right hemisphere much more that the left and perceived the world more in an intuitive sense.[10] Even though this is true, I believe that many of us still use the right hemisphere more than we realize. For example, when we have a "feeling" about something, when we have a "hunch" about something, it is probably our right hemisphere that is giving us some feedback. We mentioned earlier that the right hemisphere is better at recognizing faces than the left hemisphere. This, of course, was a great aid to survival, since being able to recognize quickly a friendly face or an unfriendly face could be a matter of life or death. This is still an extremely important matter today, and I am of the opinion that when we have a "feeling" about a certain thing or person, we should carefully consider the matter or reevaluate our relationship with the person.

The upshot of this role of the right hemisphere in modern life is that it can and does still play a significant role in our lives. I personally have come to pay more attention to my "feelings" and "hunches" than I did in the past, for in the past I did not pay too much attention to such matters and much of the time I regretted not having done so. A person can check on his "feelings" or "hunches" by simply heeding some of them in matters that are not critical and taking a mental check on the results of having heeded some of these signals. In my own case, if I have a "feeling" about a certain person that throws up a red flag, I am very cautious around the person in what I say or do. On the other hand, we may have a desire to do something or take a certain course of action and have a good feeling about it and a "feeling" that it is the right thing to do. I personally think that modern man should pay considerable attention to his intuitive thought processes, since they were formed and developed in the furnace of experience and evolution no less than the analytic thought processes. Of course, we in the West place more status upon analytic thinking, since it is used more in science and academic endeavors. But even here it may be that scientists and researchers, especially in putting forth hypotheses, act more on "hunches" and "feelings" than they like to think or admit. Perhaps the reason why we have the expression "woman's intuition" is that woman may act more on feelings or that they admit doing so more than men do. At any rate, for modern society to ignore the promptings of the right hemisphere is to waste a vital part of the brain's function.

One of the chief skills that our marvelous brain and our intelligence allow us to perform is speech and the use of language. How did mankind develop the power of speech? It must have come about over a long period of time. Blakemore believes that the upright posture was instrumental in man's learning to speak. Other factors include the grip used in swinging in the trees that helped make hands fit for precision work at a later time and, of course, the increase in the size of the brain during all this time. The natural inclination to gesture, yell, or make some noise to attract attention or alert someone—all must have contributed to the gradual use of speech.[11] Sagan cites the fact that man was often at the mercy of predators, which forced him to cooperate, and this must have involved some vocal calls. Also, the hunt demanded cooperation as well, and if one is trying to get an animal or a herd of animals to go in a certain direction, calls and yells must have played an important role in the process.[12] Sagan also points out that the large amount of brain area committed to the fingers, mouth, and organs of speech makes possible our manual dexterity and speech.[13] These factors, working together with many others, all have contributed to the intelligence of mankind, for we can well imagine that developing manual dexterity and speech, for example, must have made significant alternations in the connections and synapses in the brain itself.

Taylor believes that intelligence had its origin in thinking that was used to guide behavior in our remote ancestors. He writes: "What the study of the child suggests is that the transition from feeling to thinking was made not in a single leap but through a series of relatively small steps. In particular, it suggests that the first step was the evolution of imagination, followed by the capacity to produce images in regular sequences, followed in turn by the ability to examine these sequences and modify them in creative ways. As they do in the developing child, the emergence of these abilities would have permitted first a single-step, intuitive kind of thinking, then true thinking, and finally abstract reasoning."[14] Furthermore, Taylor believes that thinking had its origin in the development and use of language. He uses the child's use of language and later thinking as a model of how our ancestors developed in their evolution.[15] If this is correct, we can follow the process rather well by saying that intelligence, language, and thinking are inextricably intertwined in their development and in their processes and use. But we must say that without sufficient intelligence, language and thinking would not be possible.

Hunt distinguishes between natural reasoning and reasoning according to the laws of logic. Most of us, he writes, reason somewhat like a child messing around with a new toy. We frequently "mess around" with some problem, idea, or concept and the solution may come in some intuitive insight and only later do we try to put the matter in some logical form. Hunt quotes Richard Courant as saying that some mathematicians do this and later construct a proof that makes it appear as though they had reached the conclusion through logical steps.[16] Hunt says that cognitive scientists have come up with two words that best fit or describe naturalistic reasoning. The two words are *plausible* and *probable*: "For in contrast to logical reasoning, natural reasoning proceeds by steps that are credible but not rigorous, and arrives at conclusions that are likely but not certain." This kind of thinking was forged in the furnace of evolution.[17] In short, this kind of natural reasoning was the kind of reasoning that allowed our ancestors to survive the rigors of the kind of life they led. Again I believe we are led to recognize the important part that the right hemisphere played in the early life of the human race, for it was not sitting down and using logical reasoning that made it possible for them to survive in the jungle, but rather the kind of intuitive, fast, almost instinctive thinking that permits one to survive. We still use this kind of thinking all the time. Whenever we are driving and have an emergency, we don't take time to reason things out logically; if we did, we might not survive the first emergency. What we do is act immediately; our brain processes the situation in an instant, and we act. Most of the time we make the right decision, but we all know that in spite of our best efforts and fastest action, accidents still happen. Many accidents are blamed on "human error," so we know that we frequently make mistakes of judgment, but we are right enough of the time so that as a race we survive. And this was the purpose of the kind of thinking our remote ancestors used to survive, and it was forged in the evolutionary furnace.

Another function directly related to intelligence is memory. Memory is generally divided into two kinds: short-term and long-term. Short-term is the kind of memory we use when we look up a telephone number and remember it long enough to dial, but couldn't do very well trying to recall it, say, an hour later. Long-term memory is the kind that allows us to remember for long periods of time. We all remember our birth date, "1492," "1776," and other information, but curiously, when we go to write a check, we often have to check our watches to make sure of the correct date. Our memories are not infallible, but in the main, they are very good and again permit us to function very well in everyday life. What happens in our brains when we commit something to memory? We don't know exactly just what takes place, but some scientists believe they have a rather good idea of what takes place when we learn and remember. Young has this to say about the matter: "Out of all the immense amount of work that has been devoted to memory, there have emerged three types of theory about the nature of the change that establishes a memory record. There might be (1) a change of standing pattern of activity, or (2) a change of some specific chemical molecules, such as the instructional molecules of RNA, or (3) a change in the pathways between neurons within the nervous system. This last is probably the main basis for stable memory records, but all three methods are worth examining."[18]

Penfield sees the gray matter of the interpretative cortex as part of a mechanism that presents interpretations of present experiences to consciousness. Part of the cortex forms a memory file to be opened either automatically or by intentional recall. Penfield also puts great importance on the hippocampi as regards memory and recall.[19] Taylor also finds the hippocampus important in memory, especially in spatial arrangements. This observation was made from experiements on rats in cages and in mazes selecting the right arms for rewards.[20]

Blakemore is concerned with what he calls "the grammar of remembrance." On this point he writes: "From this point of view, the most compelling theory of memory is the claim that remembering might consist of the synthesis of specific molecules in the brain, the structure of each molecule representing a remembered event. This hypothesis is so powerful because it not only describes a possible physical substrate for memory (the synthesized molecule), but also embodies the nature of the code by which the information might be stored (the sequence of compounds in the molecule or its specific shape)."[21] Essentially the codes by which information and messages are transmitted in the brain and nervous system remain a mystery, but it is believed or perhaps even known that DNA and other chemicals are used in the process, along with connections and components in the brain. One of the great advantages of the human brain is that it can store vast amounts of information and get at it. Think of recalling at will dates that you know or places that you know or persons that you know. If we know something, a friend may ask us a question and if we don't have a temporary block, we can come up with the answer immediately, even though at the time our "mind" may be a thousand miles from the question we were asked.

One of the most striking characteristics of the brain to Hunt is the redundancy of its parts and circuits. This allows cells or groups of cells anywhere in the brain to be in contact with other cells in any part of the brain. Hunt believes that there is an excess of parts and circuits that would be wasteful in business, but a triumph of good design in evolutionary terms.[22] This redundancy accounts, in part, at least, for the vast amount of information, memories, feelings, and so many past experiences that we are able to store in our brains. Although the brain can retain a prodigious amount of information, some neuroscientists belive that the chief function of the brain is to filter out much of the irrelevant information and knowledge that otherwise would overwhelm us in our day-to-day experiences.[23] Without some such mechanism, our brain might from time to time become overloaded and have to shut down for a time to "cool off."

Nothing has been said in this chapter so far about intelligence, wisdom, learning, memory, and so on relating to religion. But those who are familiar with the Judeo-Christian tradition know that learning, knowledge, and scholarship have played a dominant role in the Jewish and Christian religions. I need only mention the books of Job, Psalms, Proverbs, and Ecclesiastes in the Old Testament and the Gospels and the letters of Paul in the New Testament to illustrate the point that learning and scholarship have been an integral part of these religions. But without the marvelous human brain and the intelligence thus made possible, none of the accomplishments in religion or science would have been achieved.

Chapter Five
BUILT-IN CONFLICTS

As I wrote earlier in this work, I accept the basic idea of evolution. It makes sense to me that animals and mankind have evolved over a long period of time, and far from detracting from the dignity of man, I believe that this actually adds to the dignity, worth, and importance of mankind. In short, God did not do any hurry-up job on us, but rather was willing to take millions of years to bring us to our present state of being. But this is not the scientific viewpoint of evolution, which is summed up by Taylor: "The advantage of analyzing the evolution of thought in terms of a series of distinct steps is that it reduces the complexity of the problem; if the way in which each step occurred can be understood individually, the problem as a whole can be solved. In attempting to understand these steps, however, it is important not to attribute any foresight to evolution. The adaptive significance of each step must be explained in terms of its immediate value in the struggle for survival, not in terms of its potential to support a process that has not yet evolved."[1]

The problem for me with such a statement is that I cannot imagine an indifferent, blind, perhaps even nonexistent force bringing into being a world such as we presently know. Nor can I imagine evolution not having foresight, for if it did or does not, then we are just one big accident. And that I cannot accept. It seems much simpler and much more sensible to swing Occam's razor without much restraint and cut away most of the verbiage and merely say that God created the heavens and the earth and mankind and other forms of life on earth. But this statement many scientists cannot accept.

Academically speaking, it seems that we are either oriented toward religion or oriented toward science. It appears very difficult for many people to come to grips with both disciplines. I remember that it was quite a struggle for me to do so. I was first oriented toward religion, and when I first began taking courses oriented toward science, I had the feeling that the professors were saying, "There is but one God, and scientific method is his name." If one is oriented toward one viewpoint, it seems that there is almost an inherent resistance or even hostility toward the other viewpoint, and such resistances and antagonisms are not easily overcome, even if one has the desire to do so. Nevertheless, many people do overcome such obstacles in the way of their intellectual development, and I believe that they come out stronger and more secure from having gone through the inner struggle. So I would like to make a modest proposal: that since it appears that both religion and science are here to stay, we come to grips with the other point of view than the one we presently espouse and make peace with one another. In the long run, I think we must do so.

But turning back to the idea of evolution, Beaber considers the brain one of life's greatest gifts.[2] Beaber also wonders what our brains could have been doing those long years before we used language, culture, and the other developments that came with civilization. I belive that because of the kind of lives they led, our remote ancestors used most of their brain power just to survive, for it was no easy task to find a daily supply of food, water, and shelter and outwitting those animals that preyed on them was certainly no easy task either. So the hunt, the chase, and the other neccessary tasks of survival must have been almost a full-time job at a certain stage of human development. So in the early stages of evolution the human brain was occupied with many things that could mean life or death, some of which may or may not concern us today.

The evolutionary process may be seen in the human brain. Hampden-Turner says that in reality the human brain is three brains. He breaks the divisions down as follows. First is the reptilian, which is the oldest part of the brain and consists of the matrix of the brain stem, the midbrain, the basal ganglia, and part of the hypothalamus and reticular activating system. This part of the brain is said to be a slave to precedent and is responsible, so some claim, for human tendencies toward ritual and for many of our customs and habits. The attorney's penchant for following precedents is sometimes cited as an example. The old mammalian brain is composed of the limbic system and is the seat of emotions and controls our autonomic nervous system. The newest part of the brain is the neocortex. This is the convoluted mass that covers the top of the brain and is the thinking part of the brain. It is the part that has made us human beings.[3] But the brain itself is hard to explain by evolution alone. Jastrow notes: "Among the organs of the human body, none is more difficult than the brain to explain by evolution. The powers that reside in the brain make man a different animal from all other animals. Yet this subtly wired mass of grey matter evolved as all other organs of the body evolved, by the action of natural selection in circumstances where greater brain power enhanced the prospects for survival."[4]

The eye presented a problem to Darwin in explaining the human brain and eye through the evolutionary process. It appears to be designed, and as the argument ran, you cannot have a design without a designer.[5] This is, of course, one of the main arguments against evolution. To believe that evolution and chance working alone somehow together could produce such marvelous instruments as the brain and the eye strains one's credulity to the utmost. This is why to me it makes more sense to believe that God, working through some such process as evolution, made the human brain and the eye and so on. Also when we realize that the older parts of the brain evolved between 100 million and 300 million years ago, we have an idea of the tremendous amount of time it has taken to develop the mass of neurons that we call our brain.[6] Hampden-Turner sees the old brain as responsible for such things as dominance, submission, sexual courtship and display, rituals, and mass migration. The limbic system is concerned with defending territory, hunting, hoarding, bonding, nesting, greeting, flocking, and playing. The latest addition to the brain, the cortex, is more adept at learning to cope and adapt.[7]

The many connections between the older parts of the brain and the new parts make it very likely that most of our ideas and thoughts are emotionally toned and that our emotions soon find ideas and thoughts to express themselves.[8] These connections between the old and new brain may be a physiological basis for rationalization. Most of us know that when a political candidate is giving a speech, he must not only express ideas clearly, but he must do so with enthusiasm and emotional conviction. In short, in a speech we accept the fact that emotions and ideas go together; however, this does depend somewhat upon the type of speech. For example, we would not expect a paper being read at a scientific convention to have the same fire and fervor as a presidential campaign speech. Nevertheless, even a scientific paper involves some vested interest on the part of the individual presenting the paper and we would expect that the presentation would not be completely devoid of emotion.

One of the distinctive featues of the human brain is that as newer parts of the brain evolved, the older parts were not discarded but retained, so that the cortex folds around and covers the older parts such as the brain stem and the limbic system. So what we have, in a sense, is three brains acting in cooperation with one another, but as we shall see later in this chapter, the cooperation is not always complete. One of the functions of the cortex is to control the older parts of the brain. Young comments upon this point:

> There is a continual interplay between the tendencies to action that spring from the lower centres and the restraint imposed by the learned responses of the cortex. Stimulation in the midbrain region will make one monkey attack another—but only if it is a lower member of the hierarchy. A remarkable example of inhibition of aggressive behavior is given by Delgado, who inserted electrodes into the brain of a bull and gave the bull-fighter a remote-control stimulator with which to defend himself. In man the basal (orbital) zone of the frontal cortex is an especially important inhibitor of these lower regions. People in whom it is damaged show a generalized disinhibition and changes in affect. They may show lack of self-control, violent emotional outbursts and in fact their character seems quite changed.[9]

Yet these older parts of the brain have not been completely overpowered. They remain in their old positions and do not have the same undisputed control over the body they once did, but still exercise considerable influence. Jastrow believes that these older parts of our brain continue to operate in accordance with a set of programs that go back to the mammals on the forest floor and to the reptiles.[10]

It is interesting and somewhat amusing for us humans to see animals in the courting stage. Some male birds, for example will spread their wings and do a sort of dance around the female. Other birds will do a strut or flash the colors of their tails, as a peacock does. But we humans are not exempt from certain primitive courtship patterns. Indeed, much of our courtship behavior may come from the more primitive parts of the brain. Petersen quotes the anthropologist Givens on this point: "Wooing cues are prewired in our brains. They come not from the rational part of the mind, the neo-cortex, but from the more-primitive areas of the brain, the mammalian limbic system and the reptilian core. It's as though we had a live-in serpent and a resident furry animal in our brain case making all the important decisions."[11] Human courtship, says Givens, consists of four phases. The attention phase is first and says "I am here" and "I am harmless." The second phase is the recognition phase, in which we read responses to signals, making interpretations of responses whether to move closer or to move in some other direction. Speech is the third phase, and this is said to be the most difficult of all, since many are eliminated from consideration from the very first words. The danger here is said to be that we give too much information about ourselves too soon—information about education, social status, or other personal information. The fourth and last phase is the touching phase, in which "courting takes a quantum leap toward intimacy, evoking primal responses that predate hearing, sight, and smell by millions of years."[12]

Many scientists eschew the idea of instinct in human behavior, but Restak believes it plays a prominent role in our everyday lives. He mentions our "predisposition to compulsive and ritualistic behavior, a propensity to seek and follow precedent as in legal and other matters, and a natural tendency to imitation."[13] But not all of these stereotyped instinctive behaviors are beneficial or acceptable in modern society. Restak quotes Wilson as saying: "Soon we may have to pick and choose among the emotional guides we have inherited and determine those that should be followed and those that should be sublimated or redirected, so that our behavorial patterns will both conform with biological principles and foster the growth of the human spirit."[14]

We spoke earlier of the new brain, the cortex, growing over and covering the old brain. One of the unfortunate consequences of having both the old and new brain is that they sometimes come into conflict with each other. The old brain is largely concerned with survival, and this is often survival as our remote ancestors had to contend with. On the other hand, the cortex is largely concerned with reason, thinking, and the like. Usually the two old and new brains cooperate, but as Jastrow notes, they sometimes come into conflict with each other: "In man, the cerebral cortex, or new brain, is usually master over the old brain; its instructions can override the strongest instincts toward eating, procreation or flight from danger. But the reptile and the old mammal still lie within us; sometimes they work with the highest centers of the brain, and sometimes against them; and now and then, when there is competition between the two mentalities, and the discipline of reason momentarily weakens, they spring out and take command."[15] This built-in capacity for internal conflict is also noted by Sagan, who uses the analogy of the two horses in Plato's *Phaedrus* that pull in different directions, with the charioteer barely in control.[16] This tension and constant pulling in different directions could certainly account for a great deal of the anxiety, fear, and uncertainty that many human beings experience. Indeed, it could account for some of our more serious mental illnesses, when the pulling and tugging inherent in the makeup of our brains comes to a standoff and neither the old brain nor the new brain will relent. This could make the ensuing tension and anxiety well nigh intolerable for the unfortunate individual.

Taylor also comments upon this propensity for internal conflict that we all experience from time to time. He writes: "There is another consideration. Our brains, as we have seen, comprise two main systems: the mid-brain concerned with meeting our immediate needs and the cortex which looks ahead and considers the consequences of impulsive action, including its effect on other people and their possible reactions toward us. In short, as everyone knows, there is a continuous conflict between 'reason' and 'emotion.' Thus behavioral outcomes are dependent upon a conflict which can easily swing either way, and which is highly responsive to chance external influences."[17] David A. Taylor also has observed that reason and emotion take place in different parts of the brain and were developed at different evolutionary stages. This helps us understand how conflicts between thought and emotion can so easily arise.[18]

Besides the matter of the three brains trying to operate as one, and most of the time doing very well at it, we have the matter of the two hemispheres, which could very well give us four brains or combinations working together. The left brain is said to be "dominant" in most of us. Now it seems to me that right there is another possible source of conflict, for the right brain can think and operate on its own in some matters and it may be the chief source of human creativity. Now if the right brain is aware of the left brain being called the "dominant" hemisphere, the right brain may think that its contribution to our welfare and achievements is just as important as that of the left hemisphere and resent being, at least by implication, called the inferior hemisphere. Perhaps a better word to decribe the status of the right hemisphere in most of us, by implication, might be *submissive.* But there again we have the possibility for more conflict, since to be at once dominant and submissive is a major problem with many people. And as a matter of fact, we know that all of us can at times be dominant in some situations and submissive in others. Take, for example, the man who goes to work and sometimes must be submissive to the wishes of his superiors, but when he comes home from work he can be dominant, even though in a benign manner. But few of us like to be considered characteristically submissive, while most of us would just as well not be considered domineering. I am writing these things to try in some small way to present the right hemisphere's reaction to being called by implication the inferior hemisphere. So I propose that instead of referring to the left hemisphere as the dominant hemisphere, we describe ourselves as being either left-hemisphere or as right-hemisphere "oriented." This may seem to be a minute matter, but if there is any way we human beings can eliminate some of our internal conflicts, we should certainly do so. Our very survival as a race and civilization may depend upon our taking every opportunity to reduce internal conflict, since internal conflicts often have such dire external consequences.

But the relationship of the two hemispheres may not always have been the way we regard it today. One of the most intriguing books that I have read in connection with this project is Julian Jaynes's *The Origin of Consciousness in the Breakdown of the Bicameral Mind*. Hampden-Turner does such an excellent job of summing up succinctly Jaynes's position that I would like to quote him at length:

> Jaynes believes that during a crucial period in our evolution, at the very time that language was being acquired by the left hemisphere, the right temporal lobe was preempted for the issuance of god-like commandments, across the thin anterior commissure that joins the two temporal lobes like a private corpus callosum. Auditory commands would have been the most economic code for getting elaborate information processing through so small a channel.
>
> When this hallucinatory area was stimulated by an electric current in recent experiments by Wilder Penfield, subjects would hear voices (and sometimes have visions) addressing them. One typical subject explained, "That man's voice again! My father's . . . and it frightens me!" Others heard voices to the accompaniment of music, chanting or singing, which would criticize, advise, command, but they were consistently other than the hearer, often a dead relative or friend. We also know that while the right hemisphere cannot speak, or barely so, it can comprehend and interpret quite complicated instructions. Patients with strokes in their left hemisphere can obey their doctors or researchers in detail.[19]

Jaynes believes that at a certain stage in our evolution, our remote ancestors were not conscious in the same way we are today. Consequently, their decisions were made largely by the gods who ruled them and spoke to them through the right hemisphere. Jaynes believes that the *Illiad* is an example of a time when human beings were not yet conscious, but were ruled by their gods, who gave commands through the right hemisphere. An example Jaynes gives from the *Illiad* is when Achilles reminds Agamemnon of how he robbed him of his mistress, but it really wasn't Achilles who did it, since he was commanded to do so by

Zeus. And Agamemnon accepts this as a legitimate explanation, since he also was under the command of his gods.[20] The fact that they were both military men may also have had something to do with the ready acceptance by Agamemnon, since military men know what it is to "be under orders."

What could precipitate the hearing of voices in our ancestors, who were not, as Jaynes states, as yet conscious as we understand the term? Jaynes believes that it was stress and that the preconscious people had a much lower threshold for stress than we do today. So any change in behavior or environment or anything that required making a difficult decision could trigger the voices giving advice or commands or criticizing.[21]

What about today? Does the right brain have any special significance concerning religion, worship, or God? Some think that it does. For example, Dr. James Ashbrook, a professor at a United Methodist seminary, Garrett-Evangelical Theological Seminary, believes that the use of the right brain in worship allows a holistic approach to worship or prayer, while the left-brain approach to worship promotes a more formal and logical approach to worship. Storytelling, such as Jesus employed in the parables, lends itself more to right-brain worship then preaching logically correct sermons.[22] So the ideal in worship is to engage the whole brain and not neglect either reason or emotion and feeling. And it may be that when God communicates with us, he frequently does so through the right hemisphere, which might in the long run make the right hemisphere the "dominant" hemisphere.

Chapter Six
INTERNAL CONFLICTS REVISITED

Internal conflicts are a fact of life. We read and see evidence of them everywhere and most especially within ourselves. Some of the greatest literature has centered about the theme of internal conflicts. Hamlet, which some have called the perfect work of art, has as its central theme Hamlet's internal conflict. He knows his uncle killed his father and married his mother. He also knows that he should kill his uncle and avenge his father's death, but he cannot bring himself to kill his uncle. Why can't he do what he believes he should do? Various solutions have been offered, but the thing that interests us here is that the theme of the play *Hamlet* is based on an internal conflict. Internal conflict is no stranger to the New Testament writers either. Internal conflicts are well illustrated in the writings of the apostle Paul. Basic to all his conflicts and to ours as well is the statement in Roman 7:15: "I do not understand my own actions. For I do not do what I want, but I do the very thing I hate." A psychoanalyst reading such a statement might see the seeds of self-destruction. But after our journey through the brain and mind, we see the conflict between the old brain and the new brain.

When we consider the kind of lives our remote ancestors led, we can be more understanding of the old brain and its desires, fears, and motivations. Imagine how it would be to live from day to day without any store of food. Each day's food had to be caught and killed that very day, much as the lions do in parts of Africa today or a hawk must do in our nation to survive. Furthermore, our remote ancestors were constantly in fear that some wild beast would catch them and kill them and perhaps even eat them alive. Then they had to contend with fellow human beings as enemies where the law was often kill or be killed, and we must remember that this may have been long before the cerebral cortex had evolved. So the only guide these ancestors of ours had was to survive, and had they not done so, we would not be here today. At the time we are speaking about, there was no Golden Rule, probably not much altruism, possibly just one's loyalty to one's immediate family or group and, above all, the need to survive. And it seems that much of that old brain behavior is still very much a part of us. If that sounds too strong, remember some of the things that we do to each other. We still rob, steal, kill, kidnap, rape, declare and pursue wars, cheat, lie, deceive, trick one another, and in general sometimes act as if we were still living back in remote times. Of course, we do many good and fine things as well, but right here we are concerned with the bad things we do, to illustrate Paul's statement that "I do the very thing I hate." And we should remember that these words were written by a man who above all else wanted to please God and do right. In order to get a good understanding of Paul's plight, one should read the seventh chapter of Roman and then read the thirteenth chapter of Corinthians. To do so gives us more of a full range of the personality of Paul the Apostle.

It seems that such conflicts are built in and are a part of our human nature and much that the old brain does for us is obsolete, but much that it does for us is necessary to survive. Our ability to act instantaneously, as when driving a car, must be attributed to the old brain, because thinking takes too much time in some emergencies; we must simply act and then think later, if need be. In many situations, what is needed is action not thinking. So then I attribute Paul's conflict or conflicts as stated above to a large extent to the conflict inherent between the old brain and the new brain. It is as though the old brain has one set of values and the new brain has another set of values and sometimes the two brains are quite adamant in maintaining their respective positions and "we" are caught in the middle and may suffer extreme anxiety, tension, guilt, uncertainty, or the inability to act, as in the case of Hamlet. It seems to me that this conflict could be the cause of some suicides in which a person, acting under the urging of the old brain, does something or a series of somethings that arouse indignation, horror, and revulsion in the new brain and a fight to the death ensues. In short, part of us is enraged against another part of us, so we have internal civil war. This most likely would occur in those individuals with a very "sensitive conscience."

In Romans 7:21–25, Paul gives us another example of an internal conflict: "So I find it to be a law that when I want to do right, evil lies close at hand. For I delight in the law of God, in my inmost self, but I see in my members another law at war with the law of my mind and making me captive to the law of sin which dwells in my members. Wretched man that I am! Who will deliver me from this body of death? Thanks be to God through Jesus Christ our Lord! So then, I of myself serve the law of God with my mind, but with my flesh I serve the law of sin." With our present understanding of the human brain, we can paraphrase Paul's last sentence above to read: So then, with my new brain I serve the law of God, but with my old brain I still serve the law of sin. The seventh chapter of

Romans is generally taken to be autobiographical, and so it is not some hypothetical situation that Paul is discussing. He is talking about real internal conflicts with which he must deal on an ongoing basis. He wants to do right, but finds that evil is always close at hand. Part of him delights in the law of God, but part of him wars against God's law, and he winds up a captive to the law of sin. He finds himself hopelessly entangled in this internal war or conflict, as we call it. Only Jesus Christ can extricate Paul from his wretched situation and restore him to wholeness again. Paul's summary is that part of him, his mind, serves the law of God and part of him, his flesh, serves the law of sin.

Again we see this internal conflict in terms of the struggle between the old brain, with its primitive demands and needs, and the new brain, with its inclination toward reason, logic, and morals. The old brain cannot give up its need to help ensure survival of the individual; the new brain cannot give up its proclivity to be social, to consider others, and to "treat others as we would like to be treated." On the surface, this may seem like an incompatible situation, and in some individuals it undoubtedly is that way. Surely the conflict between the old and new brains must account for much of the guilt that we human beings experience. It is something like a tug-of-war except in this case, instead of one side going back and one going forward, both sides go back. That is, in some internal struggles, both the old and the new brains become more adamant in regard to their positions. The result is often personal guilt, anxiety, fear, and an extreme loss of confidence. It may be that the individual acting under the initiatives of the old brain does something or perhaps just thinks something that outrages the new brain and the struggle ensues. Sometimes the situation actually becomes incompatible and there is a severe emotional illness as a result. But often the individual, through some act such as changing vocations, changing his life-style, having a religious experience, undergoing psychotherapy, or giving up something, can bring about a compromise and produce a truce between the warring factions. It may be that the person takes a religious solution, as the apostle Paul did. In the Christian culture, this usually takes the form of repentance, baptism, joining a church, and changing one's life-style. If this does not actually resolve the internal conflict, it at least usually neutralizes it, and the individual can then go on about his new life in a comfortable manner.

In Galatians 5:16–17, Paul puts the conflict in a different perspective when he writes: "But I say, walk by the Spirit, and do not gratify the desires of the flesh. For the desires of the flesh are against the Spirit, and the desires of the Spirit are against the flesh; for these are opposed to each other, to prevent you from doing what you would." Here the conflict is found between the human flesh and the Spirit and since the word *Spirit* is capitalized, it refers to the Spirit of Christ or the Holy Spirit—that is, God's indwelling Spirit. Where does God's Spirit dwell in the human being? We have said that it must be in the right hemisphere of the brain. So once again we have internal conflict between the old and new brain, with the added dimension of the Spirit allied with the new brain. When we realize that the old brain has millions of years behind it and that we are so oriented toward the new brain in modern society, even going so far as to deny that some of the old brain emotions are really ours, we can realize the intensity of the struggle that can ensue. So it may be that the Spirit of Christ is necessary as an ally for the new brain to win the struggle against the old brain with some of its primitive drives.

Even Jesus did not escape the internal struggle between the old and new brain, as Mark 14:32–36 (GNB) tells us:

> They came to a place called Gethsemane, and Jesus said to his disciples, "Sit here while I pray." He took Peter, James, and John with him. Distress and anguish came over him, and he said to them, "The sorrow in my heart is so great that it almost crushes me. Stay here and keep watch."
>
> He went a little farther on, threw himself on the ground, and prayed that, if possible, he might not have to go through that time of suffering. "Father," he prayed "my Father! All things are possible for you. Take this cup of suffering away from me. Yet not what I want, but what you want."

So we find Jesus struggling with the internal conflict of wanting to serve God and do his will, on the one hand, and on the other hand not wanting to suffer and die. This may not be purely an example of the old brain and new brain conflict, and yet it certainly has the elements of being so. The old brain says to survive and live—that is the thing that matters most. The new brain, and perhaps the right hemisphere in particular, says that the right thing to do is the will of God—do what the Father requires of you, to do otherwise is wrong. That is why Jesus can say that he is almost crushed. The anxiety, the pulling, and tugging of the internal conflict have taken their toll on him. This struggle had undoubtedly been going on inside Jesus for some time and may have started as early as the temptations. So what we have reported for us by Mark as quoted above is the culmination of the struggle. And the new brain won only with great effort and struggle and suffering.

Jesus knew mankind very well, as John 2:23–25 tells us; "Now when he was in Jerusalem at the Passover feast, many believed in his name when they saw the signs which he did; but Jesus did not trust himself to them, because he knew all men and needed no one to bear witness of man for he himself knew what was in man." Since Jesus would not trust himself to the people in question and he knew what was in man, John must have been referring to those evil qualities in man of which Jesus was aware. Jesus was very cautious in his relationships with some of his fellowmen. It is true that he was quite open with some of his contemporaries, but still cautious around others. What did Jesus see in man that made him, as John says, not trust himself to them? Well, it must not have been the desirable, noble qualities, for these would not have deterred him from trusting himself to them. So it must have been some bad or evil qualities that Jesus found in man that made him cautious around some of them. Mark 7:20–23 tells us what these undesirable qualities were that Jesus found in some of his fellow men: "And he said, 'What comes out of a man is what defiles a man. For from within, out of the heart of man, come evil thoughts, fornication, theft, murder, adultery, coveting, wickedness, deceit, licentiousness, envy, slander, pride, foolishness. All these evil things come from within, and they defile a man.' "

Of the things Jesus listed, we may discard some in his relations with his fellowmen and list the ones that he was concerned about in regard to entrusting himself to the men that he would not trust himself to. So we may take from the list as those qualities that made Jesus cautious the following: evil thoughts, theft, murder, coveting, wickedness, deceit, envy, slander, pride, and foolishness. This still constitutes a lengthy list of undesirable traits, and from that list I pick murder, deceit, and slander as the three that concerned him most in regard to his own welfare. Of these I further limit Jesus' concern to murder and deceit, for in the long run it was these two characteristics that his enemies used to bring him down. These two, murder and deceit, we may suppose were used quite frequently by our remote ancestors in their struggle for survival. Having been used so often, murder and deceit are surely tied rather closely to the old brain. We may be sure that in order to survive, many of our ancestors resorted to murder and deceit; and at a certain stage of our evolution, they may have been acceptable tools in surviving. If you could trick a person, so much the better, and it seems that in some circles today the idea is still prevalent. In wartime, for example, the tactic of trying to make an enemy think you are going to launch an attack in one area, but making the actual attack in another area is just plain good strategy. So for Jesus to entrust himself to people who employed deceit and murder would have been just plain poor strategy and judgment on his part. But when it comes to dealing with our fellowmen, we are still too often led by the old brain, which views most people as enemies or as potential enemies, and that is one reason why murder and deceit are still rife in modern society. It is time to proscribe such behavior as inappropriate in modern society.

Finally, there is one statement of Jesus that has long intrigued me and made me wonder just what he meant by it. In John 16:12, Jesus says to his disciples, "I have yet many things to say to you, but you cannot bear them now." What did he have to tell them that they couldn't bear at that time? Surely, one of the things would have been the future. To know the trials, the suffering, the hardships, disappointments, and death that awaited them surely would have been too much for them at the time. Indeed, how many of us would want to know all that the future holds for us? If we knew just how much we had to endure, the difficulties that awaited us in the future, it might well be too much for us to bear at one time. So this was surely one thing that Jesus could have told them, but did not.

Another thing that I think that Jesus might have told the disciples was this: There is no devil. After having gone through our study of the brain and mind, it seems to me that we may very well say with confidence, "We have met the devil, and he is part of us." Too long the devil has taken a bum rap from the human race. He has been our excuse; whenever we have done something wrong or sinned, we may say Satan or the devil tempted us. Some of you may remember the comedian Flip Wilson. One of his favorite characters was Geraldine. Geraldine's favorite expression was "The devil made me do it." If she was by a store window and saw an expensive dress she liked, she would just go into the store to look, but the devil made her buy the dress. For far too long now, we have used the devil in just this way—to escape responsibility for our own conduct and actions. Many Christians have taken advantage of the concept of the devil in just this way. So instead of saying, "I did such and such and I was wrong," the tendency is to say, "The devil was after me, and he finally won the battle and I gave in," or offer some such excuse involving the devil. It is worth noting that in the seventh chapter of Romans, Paul says nothing about the devil causing him to do the very things he hates; no, he takes full responsibility himself, using the pronoun *I*. When Jesus spoke to the disciples and made the statement that they could not bear to hear some things at that time, he was right. But now I feel that the time has come for us as mature people and, more particularly, as mature Christians to renounce such excuses and take responsibility for our own actions, as painful as that may be on occasions.

Many Christians will be greatly offended by the above statements and possibly have some unkind thing to say about such a person who would write such things. But I am prepared for that and take full responsibility for my statements and certainly will not blame the devil and say that he put me up to it, although I am sure that many of my readers will be convinced that the devil did put me up to it. But our old-brain behavior is so much a part of us that we hardly notice it at times. Surely it is time that we come to grips with where we came from and how our ancestors managed to survive so that we might be here today. We should not be ashamed of what old-brain mentality we have; after all, it helped us get this far in human history and it surely was one of the ways God used for our remote ancestors to survive. We don't like to think about the things that our remote ancestors had to do to survive, let alone admit that some of those things still survive in us today. But all one has to do is pick up any newspaper or listen to a news broadcast to realize how much of the primitive life still survives in us today. Violence, robbing, cheating, deceit, and stealing are all still very much a part of our modern society. So it is time we gave the devil his due and admit that within us lies much of the evil that is so much a part of modern life. The disciples could not bear such tidings; today I think many of us can and indeed must.

Some of the things I have written here may sound new and strange to some readers, but I believe that the sooner we realize that the conflicts between the old and the new brains are the source of much of our suffering, anxiety, fear, tension, and the like, the sooner we may have opportunity to deal constructively with our internal conflicts. The survival of the human race may depend upon our doing so.

Chapter Seven
BRAIN, RELIGION, AND SCIENCE

Much of our activity is directed toward survival. This is just as true of animals as well; as nature, or God, has provided for ways in which life can go on. Not all species do survive, as we know, but many do survive and some, such as the crafty coyote, have survived in spite of man's best efforts to exterminate them. Part of the reason for survival of the human species is that our brains are programmed for various functions. For example, the frontal lobe of the cortex is thought to be programmed for certain feelings and needs. Glasser has this to say about the frontal lobe: "The size of the frontal lobe, much smaller than ours in the higher primates, is one of the obvious differences between our brain and theirs. If our huge, recently developed frontal lobe is the seat of the general needs, that is, the strong needs such as love and worth discussed in chapter 1, it would follow that severe injury to the frontal lobe would produce a person who needs very little. In practice this is exactly what happens."[1]

The right hemisphere is involved in processing emotional material. Restak cites experiments that indicate that emotionally loaded questions are dealt with in the right hemisphere; some of this emotionally loaded material gets to the left hemisphere via the corpus callosum but some evidently is kept in the right hemisphere for futher processing or perhaps is made "unconscious."[2] Restak goes on to state that trying to understand the brain involves some faith. He notes: "Behind every attempt to understand the brain lies an article of faith. Chomsky has gone even further and called it the modernist equivalent of a 'religious belief.' It concerns our access to the relevant brain processes that we must understand in order to explain how our brain works and, in the process achieve an integrated self-understanding. One of the articles of much traditional scientific faith assumes that we ultimately may become aware of—in a very personal and subjective way—those brain processes that are responsible for our behavior."[3]

Tart has pointed out that modern Western psychology has largely ignored the spiritual aspects of mankind and some even label the spiritual as pathological.[4] Tart goes on to say that some people may be able to drop their background in traditional psychology, but most of us would want to try to form some kind of synthesis and weave the spiritual and the scientific, and especially psychology, together in some sort of compatible relationship. His conclusion to the matter is: "So I think our job will be to bridge the spiritual and our Western, scientific side."[5] That is exactly one goal of this book, and if it makes a small contribution in that direction, I will be satisfied. Many notable individuals have done a good job of successfully bridging science and religion. One of the best known is Albert Schweitzer, who first became a theologian and philosopher and later a medical doctor. But to Schweitzer the study of the natural sciences had an unexpected bonus: "[T]he study of the natural science brought me even more than the increase of knowledge I had longed for. It was to me a spiritual experience. I had all along felt it to be psychically a danger that in the so-called humanities with which I had been concerned hitherto, there is no truth which affirms itself as self-evident, but that mere opinion can, by the way in which it deals with the subject matter, obtain recognition as true."[6]

Another example of a scientist who bridged science and religion in his own life is Wilder Penfield. The idea that the two could coalesce in one belief system came somewhat as a revelation to him: "Since every man must adopt for himself, without the help of science, his own way of life and his personal religion, I have long had my own private beliefs. What a thrill it is, then to discover that the scientist, too, can legitimately believe in the spirit."[7] So many individuals do find that religion and science can live harmoniously in the same belief system. This is not to say that these two disciplines can reach a satisfactory accommodation in the same brain. After all, there are disagreements within the disciplines themselves and that does not keep them from growing and achieving positive results. But Jaynes goes much further than this and says that much of the scientific impetus came from religious roots. On this point he writes:

> If we would understand the Scientific Revolution correctly, we should always remember that its most powerful impetus was the unremitting search for hidden divinity. As such, it is a direct descendant of the breakdown of the bicameral mind. In the late seventeenth century, to choose an obvious example, it is three English Protestants, all amateur theologians and fervently devout, who built the foundations for physics, psychology, and biology: the paranoic Issac Newton writing down God's speech in the great universal laws of celestial gravitation; the gaunt and literal John Locke knowing his Most Knowing Being in the riches of knowing experiences; and the peripatetic John Ray, an unkempt ecclesiastic out of a pulpit, joyfully limning the Word of his Creator in the perfecting of the design of animal and plant life. Without this religous motivation, science would have been mere technology, limping along on economic necessity.[8]

Some of the tender emotions that the human brain makes possible work to our advantage and certainly help in our survival. One way this especially shows up is in the way we care for our offspring, for without the love, care, concern, and willingness to do the many things it takes to rear children, we could not survive as a race. Worship is also one of the tender emotions. After speaking of worship in rather poetic words, Young has this to say about the subject: "All this may seem to be exaggerated and poetic language, but it is quite close to our concept of the general pattern for life. We have repeatedly emphasized the unity of the program. From time to time it needs to be exercised and tried out as a whole. This is what happens when we worship, whether in a quiet contemplation of a landscape or in a religious ceremony. We allow the reward systems to pour their gladness over the whole range of the cortical programs."[9] This no doubt is one reason why worship often makes us feel better: Our whole cortex has been bathed with a soothing "gladness," as Young puts it. It may be that worship when it is special for us may in some way release a chemical that in itself is relaxing, comforting, and reassuring. So we leave the sanctuary feeling that we have been in the presence of our Maker and the Creator and Sustainer of the Universe. Young goes on to say that beliefs are essential to all reasoning, but they themselves are not the result of reasoning. Beliefs come from our trusting capacity of the human brain.[10]

The coming and going of the seasons allow us to believe in an orderly universe and not in some sort of chaos. The book of Psalms stresses the orderly nature of God's universe and the dependability of the world in which we live. So we can believe that God is an orderly God who does things not at whim, but with a certain order and purpose in mind. Our brains seem to be programmed to accept such a frame of reference, and this also makes it easy for us to believe.[11] Believe in what? Well, to believe in God and, if we are Christian, to believe in Christ as the Son of God. Jaynes sees Jesus' ministry as an attempt to bring Judaism out of an external approach to an inner, conscious approach to religion and God. On this point Jaynes writes: "A full discussion here would specify how the attempted reformation of Judaism by Jesus can be construed as a necessarily new religion for conscious men rather than bicameral men. Behavior now must be changed from within the new consciousness rather than from Mosaic laws carving behavior from without. Sin and penance are now within conscious desire and conscious contrition, rather than in the external behaviors of the decalogue and the penances of temple sacrifice and community punishment. The divine kingdom to the regained is psychological not physical. It is metaphorical not literal. It is 'within' not in extenso."[12]

We find passages in the Gospels emphasizing the necessity for internal change and cleansing, such as Matthew 7:3: "Why do you see the speck that is in your brother's eye, but do not notice the log that is in your own eye?" Again in Matthew 23:26 we find: "You blind Pharisee! First cleanse the inside of the cup and of the plate, that the outside also may be clean." And there are many other such passages in the New Testament that emphasize the necessity for an internal house cleaning.

Many psychologists by the very nature of their work say about the same thing about mankind as Jesus said, but in different language. Tart finds that much of the evil in the world lies within ourselves; "We look at economics, politics, ecology, crime, and so on to find the villain, but we are the villains. Ordinary man, man who does not know himself, neurotic man, psychotic man—project their psychological inadequacies and conflicts out on the world, finding the villains out there and retaliating against them."[13] So both religion and modern science find that the source of much of the evil around us and in the world is ourselves. This is difficult for many of us to believe or sometimes even to contemplate. But again history and current events all attest to this fact. But this picture is not all bleak. For as Young points out, even this condition is subject to change: "As we learn to understand the brain better we can recognize that even if these tendencies have hereditary backgrounds they are not therefore equally and forever immutable. We are no longer apes, or paleoliths. Human nature has changed in the past and is likely to change even more rapidly in the future. Our lives have become very different from those of our ancestors and our descendents will be different from us again. Multi-parental inheritance of information will produce much more rapid change than the genetic evolution that has transformed ape into man."[14] As Restak reminds us, the kind of world we are in the prosess of creating for ourselves and those who will come after us is the product of our brains. Our subjective feelings, values, and goals are the source of much of our motivation, and as our knowledge of brain function increases, we should be able to deal more adequately with the study of consciousness in all its manifestations. Finally, for the first time in history, Restak believes, the interest of science and the interest of the humanities coincide.[15] I would just like to add that religion, working in cooperation with the sciences, will also make a significant contribution to the world that we are creating and the one that shall be created in the future. For just as our brains are programmed for various other functions, I believe that they are also programmed for religion and some of our deepest yearnings are satisfied as we worship God and fit ourselves into the universe that he has created.

NOTES

Chapter One. The Marvelous Brain

1. Colin Blakemore, *Mechanics of the Mind* (New York: Cambridge University Press, 1977), p. 184.
2. John C. Eccles, *The Understanding of the Brain* (New York: McGraw Hill Book Company, 1977), pp. 1–2.
3. Richard M. Restak, *The Brain: The Last Frontier* (New York: Doubleday and Company, Inc., 1979), p. 383.
4. Ibid., pp. 113–14.
5. David A. Taylor, *Mind* (New York: Simon and Schuster, 1982), p. 32.
6. W. Maxwell, "The Development of the Brain," in *The Brain: A Scientific American Book* (San Francisco: W. H. Freeman Company, 1979), p. 17.
7. Charles F. Stevens, "The Neuron," in *The Brain: A Scientific American Book* (San Francisco, California: W. H. Freeman Company, 1979), p. 17.
8. Walle J. Nauta and Michael Feirtag, "The Organization of the Brain," in *The Brain: A Scientific American Book* (San Francisco, California: W. H. Freeman Company, 1979), p. 48.
9. Restak, p. 367.
10. J. Z. Young, *Programs of the Brain* (New York: Oxford University Press, 1978), pp. 44–46.
11. Robert Jastrow, *The Enchanted Loom* (New York: Simon and Schuster, 1981), p. 75.
12. Wilder Penfield, *The Mystery of the Mind* (Princeton, New Jersey: Princeton University Press, 1975), p. 19.
13. Stevens, p. 15.
14. Gordon Rattray Taylor, *The Natural History of the Mind* (New York: E. P. Dutton, 1979), p. 51.
15. Jastrow, pp. 75–76.
16. Restak, pp. 29–30.
17. Taylor, Gordon Rattray, p. 43.
18. Young, p. 64.
19. Thomas L. Bennett, *Brain and Behavior* (Monterey, California: Brooks/Cole Publishing Company, 1977), p. 28.
20. Morton Hunt, *The Universe Within* (New York: Simon and Schuster, 1982), p. 211.
21. Young, p. 113.
22. Restak, p. 299.

23. Nigel Calder, *The Mind of Man* (New York: The Viking Press, 1971), p. 30.

24. Francis Leukel, *Introduction to Physiological Psychology* (Saint Louis, Missouri: The C. V. Mosley Company, 1972), p. 371.

25. Ibid., p. 300.

26. Taylor, Gordon Rattray, pp. 32–33.

27. Restak, pp. 48–49.

28. Bennett, p. 139.

29. Calder, p. 58.

30. Leukel, p. 301.

31. Jastrow, pp. 131–32.

32. Young, p. 144.

33. Taylor, David A., p. 117.

34. Jastrow, p. 134.

35. Bennett, pp. 46–47.

36. Jastrow, p. 134.

37. Young. p. 38.

38. William Glasser, *Stations of the Mind* (New York: Harper and Row, Publishers, 1981), p. 18.

39. Jastrow, p. 53.

40. Restak, p. 172.

41. Leukel, p. 103.

42. Taylor, Gordon Rattray, pp. 40–41.

43. Julian Jaynes, *The Origin of Consciousness in the Breakdown of the Bicameral Mind* (Boston: Houghton Mifflin Company, 1976), pp. 114–15.

44. Norman Geschwind, "Specializations of the Human Brain," in *The Brain: A Scientific American Book* (San Francisco, California: W. H. Freeman Company, 1979), p. 108.

45. Charles Hampden-Turner, *Maps of the Mind* (New York: Collier Books, 1981), p. 86.

46. Restak, pp. 171–72.

47. Jaynes, pp. 118–19.

48. Ibid., pp. 343–44.

49. Robert E. Orstein, *The Psychology of Consciousness* (New York: Penguin Books, 1972), p. 78.

50. Jaynes, p. 115.

51. Blakemore, pp. 1–4.

52. Restak, p. 28.

53. Ibid., pp. 35–36.

54. Jastrow, p. 62.

55. Ibid., p. 143.

56. David H. Hubel, "The Brain," in *The Brain: A Scientific American Book* (San Francisco, California: W. H. Freeman Company, 1979), p. 4.

57. Taylor, Gordon Rattray, p. 275.

Chapter Two. The Enigmatic Mind

1. Jaynes, p. 1.
2. Taylor, Gordon Rattray, pp. 316–17.
3. Hunt, p. 30.
4. Penfield, p. 80.
5. Taylor, David A., pp. 29–30.
6. Ibid., p. 216.
7. Young, p. 216.
8. Taylor, David A., p. 165.
9. Restak, p. 220.
10. Penfield, pp. 75–76.
11. Ibid., p. 61.
12. Taylor, David A., p. 37.
13. Penfield, p. 40.
14. Calder, p. 253.
15. *The Interpreter's Dictionary of the Bible,* s.v. "Mind."
16. Ibid., pp. 383–84.

Chapter Three. Consciousness

1. Blakemore, pp. 43–44.
2. Taylor, David A., p. 167.
3. Blakemore, pp. 34–35.
4. Taylor David A., p. 169.
5. Taylor Gordon Rattray, p. 76.
6. Jastrow, p. 16.
7. Ornstein, p. 33.
8. Jaynes, p. 220.
9. Blakemore, p. 37.
10. Hunt, pp. 354–55.
11. Jaynes, p. 9.
12. Penfield, p. 18.
13. Taylor, Gordon Rattray, pp. 74–75.
14. Ibid., pp. 318–19.
15. Restak, pp. 176–77.
16. Ibid., pp. 175–76.
17. Restak, pp. 285–86.
18. Taylor, Gordon Rattray, pp. 80–81.

Chapter Four. Human Intelligence

1. David Wechsler, *The Measurement and Appraisal of Adult Intelligence* (Baltimore: The Williams and Wilkins Company, 1958), p. 7.
2. Ibid., pp. 6–7.
3. Jastrow, p. 140.
4. Carl Sagan, *The Dragons of Eden* (New York: Ballantine Books, 1977), p. 108.
5. Ibid., p. 87.
6. Ibid., p. 38.
7. Ibid., p. 44.
8. Wechsler, pp. 19–20.
9. Sagan, pp. 182–83.
10. Ibid., p. 177.
11. Blakemore, pp. 146–47.
12. Sagan, p. 104.
13. Ibid., pp. 33–34.
14. Taylor, David A., p. 209.
15. Ibid., p. 211.
16. Hunt, pp. 136–37.
17. Ibid., pp. 137–38.
18. Young, p. 81.
19. Penfield, pp. 35–36.
20. Taylor, Gordon Rattray, pp. 30–31.
21. Blakemore, pp. 106–7.
22. Hunt, pp. 38–41.
23. Taylor, Gordon Rattray, pp. 234–35.

Chapter Five. Built-in Conflicts

1. Taylor, David A., p. 209.
2. "The Human Mind," *Tulsa World*, 21 August 1983.
3. Hampden-Turner, p. 80.
4. Jastrow, p. 104.
5. Ibid., pp. 96–97.
6. Ibid., p. 134.
7. Hampden-Turner, p. 80.
8. Taylor, Gordon Rattray, p. 154.
9. Young, pp. 158–59.
10. Jastrow, p. 127.
11. "Courtship Signals Are Subtle," *Tulsa World*, 29 December 1983, Sect. B 15. Reprinted by permission.
12. Ibid.

13. Restak, p. 40.
14. Ibid., p. 62.
15. Jastrow, pp. 132–33.
16. Sagan, p. 83.
17. Taylor, Gordon Rattray, p. 311.
18. Taylor, David A., p. 40.
19. Hampden-Turner, p. 92.
20. Jaynes, pp. 72–73.
21. Ibid., p. 93.
22. Tim Quinlan, "Worship and the Right Side of the Brain," *Aware*, Spring 1983.

Chapter Seven. Brain, Religion, and Science

1. William Glasser, *Stations of the Mind* (New York: Harper and Row, Publishers, 1981), p. 42.
2. Restak, pp. 285–86.
3. Ibid., p. 327.
4. Charles T. Tart, ed., *Transpersonal Psychologies* (New York: Harper and Row, Publishers, 1975), p. 5.
5. Ibid.
6. Albert Schweitzer, *Out of My Life and Thought*, trans. C. T. Campion (New York: Henry Holt and Company, 1949), p. 104.
7. Penfield, p. 85.
8. Jaynes, p. 435.
9. Young, pp. 260–61.
10. Ibid., p. 251.
11. Ibid., pp. 253–54.
12. Jaynes, pp. 318–19.
13. Tart, p. 3.
14. Young, p. 270.
15. Restak, pp. 379–80.